Breaking Down Projects

A

Comprehensive Guide

to

Work Breakdown Structures

vs Others

23 Most Powerful Methods for Success:

Unveiling the WBS Secrets

ERNEST D LEHMAN

Knowledge & Competency Series

Published by: ELU Publishing

http://www.ernestlehman.com

ISBN: 9798865833604

Second Edition: Oct 2023

Disclaimer

The references are based on the information available at the time of writing and are subject to change as new research and resources become available. Any URLs provided for each reference lead to the respective books or journals' websites where the original work can be accessed.

Please note that access to some of these resources may require a purchase or a subscription, as some articles may be behind a paywall.

While every effort has been made to ensure the accuracy of these references, the authors and publishers cannot guarantee the ongoing validity of the URLs, as website structures can change over time or content may be moved or removed.

It's recommended to use the provided DOI (Digital Object Identifier) where available, as this provides a persistent link to the electronic location of the content.

For higher education students, please ensure that you research each paper for referencing, as it might not align with the intended message of your scholastic deliverable.

Who should read this book?

Dive into the pages of this book, where my four-decade journey in construction and project management comes to life, offering a unique blend of insights and expertise. Drawing from my vast knowledge on this subject matter, from PMBOK to IPMA, this book is more than just a guide; it's a gateway to mastering project services and controls.

"In 'Breaking Down Projects' my extensive expertise in business analytics and project management comes to life, offering invaluable advice for a diverse readership.

This book is an essential guide for:

Project Managers: Regardless of experience level, project managers will find this book indispensable for crafting precise work breakdown structures (WBS) and other critical project structures, enhancing their project execution capabilities.

Team Leaders and Members: Empowering team members to align their contributions with overarching project goals, thus playing a pivotal role in the project's success.

Business Owners and Executives: A comprehensive resource, it offers business leaders an in-depth understanding of WBS and project structures, aiding in effective decision-making, resource allocation, and aligning projects with organizational objectives.

Students and Academics: As an academic cornerstone, it's a treasure trove for students and professionals keen on augmenting their project management acumen, based on PMBOK and other recognized methodologies.

Professionals in Related Fields: Those in engineering, construction, software development, marketing, and more will find this book enhances their understanding of project structures, contributing to their professional growth and success in their respective fields.

TABLE OF CONTENTS

FIGURES

TABLES

PART I – WBS STRUCTURES

CHAPTER 1

Introduction

Importance of effective project management

As a project team, you are well aware of the importance of effective project management. A project can easily fall apart without proper management, causing major delays, cost overruns, and even project failure. Effective project management is not just important for completing a project successfully, but it is also essential for achieving the project goals and objectives within the given timeline and budget. In this chapter, we will discuss the importance of effective project management using work breakdown structures (WBS) and how it can benefit the project team and the organization. I will mostly refer to a few projects in this book: The Extra-Terrestrial Project (ETP) and the Guest House Construction Project (GHP).

First and foremost, effective project management helps ensure everyone involved is on the same page. It establishes clear goals, objectives, timelines, and expectations, which helps to prevent confusion and misunderstandings. This way, every team member knows what they are responsible for and what they need to do to achieve the project's goals. This clarity helps ensure everyone works together towards the same common goal, preventing unnecessary delays or mistakes.

In project management, the Work Breakdown Structure (WBS) is a vital tool that provides a systematic and organized approach to managing and executing projects. It is essentially a visual representation of the project scope, outlining all the tasks, activities, and deliverables required to complete the project successfully. The WBS breaks down the project into manageable components, making it easier for project teams to plan, schedule, and monitor their progress. The various WBS formats that we will show in this book align with the objectives, cost, time, responsibilities, etc., and most importantly, project deliverables.

Another important benefit of effective project management is that it helps to identify potential risks and issues before they become major problems. By identifying these risks early on, the project team can take steps to mitigate them, reducing the chances of delays or cost overruns. Effective project management also helps track progress and performance, allowing the team to adjust as needed to keep the project on track. We will utilize some tools and techniques to assist the project team in managing the project "scope" with various structures that can be used during any project life cycle phase.

Using these techniques, you assist the team with effective project management to ensure the project is completed within the given timeline and budget. By establishing clear timelines and budgets, the project team can monitor progress and ensure the project stays on track. This helps prevent cost overruns and delays, which can harm the project's success.

Finally, effective project management helps ensure the project is completed to the highest possible quality. By establishing clear objectives and standards, the project team can monitor progress and ensure that the project meets these standards. This helps the project deliver the desired outcomes, which can be critical for its success.

In conclusion, effective project management is crucial for the success of any project. It helps to ensure that everyone involved in the project is on the same page, identifies potential risks and issues early on, tracks progress and performance, ensures that the project is completed within the given timeline and budget, and ensures that the project is completed to the highest possible quality. As a project team member, it is essential to understand the importance of effective project management and to work together to achieve project success.

Focus Areas of this Book

In this book we focus on the areas the work breakdown structures (WBS) and all related types of tools and techniques that will assist you in developing structures for your project team. This book relates to my work experiences and the inputs that I have encountered during my elaborate career in project management. I have been following the PMBOK guides for nearly 3 decades and found them to be quite insightful. This book uses PMBOK section 5.4.

NOTE: You'll see these notes in various sections of this book and they are suggestions and real-life experiences. I also cover key tips and lessons learnt at the end of each chapter.

So here is a visual representation of what we cover in this book.

	Project Management Process Groups				
Knowledge areas	Initiating Process Group	Planning Process Group	Executing Process Group	Monitoring and Controlling Process Group	Closing Process Group
5. Project Scope Management		5.1 Plan Scope Management 5.2 Collect Requirements 5.3 Define Scope 5.4 Create WBS		5.5 Validate Scope 5.6 Control Scope	
Source: PMBOK® Guide, 6th ed., Chapter 1, section 1.4, p. 25. (Extract)					

4. Create WBS		
Tools and techniques		
Inputs	**Tools and techniques**	**Outputs**
Project management plan	Expert judgment	Scope baseline
Scope management plan	Decomposition	Project documents updates
Project documents		Assumption log
Project scope statement		Requirements documentation
Requirements documentation		
Enterprise environmental factors		
Organizational process assets		
Source: PMBOK® Guide, 6th ed., Chapter 5, section 5.4, p. 156.		

You will find that during my elaborate career I have used references and guides from various institutions (See the Author page). In addition, during my studies I compared and analyzed the pro's and con's and utilized their advice to ensure my project successes. As an example, some institutions were more explicit on soft issues as others a few years back.

Body of Knowledge	PMI	IPA
PM PROCESSES	Integration	Project Management Integration
	Scope	Project Scope Management
	Schedule	Schedule Management
MANAGEMENT & CONTROL PROCESSES	Integration	Business Case Development
	Cost	Cost Control
	Risk	Risk Management
	Integration	Change Management
	Procurement	Procurement & Contracting Mgmt
	Schedule	Design and Engineering Mgmt
TECHNICAL DELIVERY PROCESSES	Human Resources	EPCM Management
	Human Resources	Operational Readiness Planning
	Schedule	Commissioning & Ramp Up Mgmt
	Integration	Close Out Management
	Communications	Communication & Stakeholder Mgmt
	Scope	Value Improving Practices
SUPPORT PROCESSES	Quality	Quality Management
	Integration	Document Management
	Human Resources	HR & IR Management
	Human Resources	Safety & SD Management

PMBOK vs IPA Focus Comparison

Overview of the book

The book "Breaking Down Projects: A Comprehensive Guide to Work Breakdown Structures vs Others" is a must-read for project team members who are looking to gain a deeper understanding of project management work breakdown structure (WBS) and other structures. This book is a comprehensive guide that provides a detailed overview of how to break down projects into manageable tasks and sub-tasks.

Part I of this book covers the basics of WBS and other structures, including the purpose of creating a WBS, the benefits of using a WBS, and how to create a WBS. It also provides an overview of the different types of structures that can be used in project management, including organizational breakdown structure (OBS), product breakdown structure (PBS), and work package breakdown structure (WPBS).

One of the key benefits of this book is that it provides practical examples and case studies that demonstrate how to apply WBS and other structures to real-world projects. The illustrations and case studies are designed to help project team members understand the benefits of using WBS and other structures and the potential challenges that may arise when using these structures.

Another key feature of this book is that it provides a step-by-step guide on how to create a WBS. The guide takes the reader through the process of defining the project scope, identifying the project deliverables, and breaking down the deliverables into smaller, manageable tasks. The guide also provides tips and best practices for creating a WBS that is effective and practical.

Part II of this book delves into creating and utilizing the WBS dictionary, an essential tool for enhancing project planning, communication, and execution. It offers a step-by-step approach to crafting precise element descriptions, ensuring a unified team understands project goals and scope:

the key to mitigating risks and avoiding costly delays.

As you progress, you'll gain insights into maintaining and updating the WBS dictionary with practical tips to navigate project teams' everyday challenges. This section is replete with best practices and strategies to refine your project management process, ultimately leading to heightened project success and improved team synergy.

By the conclusion of Part II, you'll grasp the significance of a WBS dictionary and possess the expertise to implement it effectively, propelling your project teams toward unparalleled project management proficiency.

Lastly, Part III gives you the tools to create a WBS and review & assess all your project structures using an approved method for guaranteed results. This section also includes a project office guideline to set up WBS structures.

Overall, "Breaking Down Projects: A Comprehensive Guide to Work Breakdown Structures vs Others" is an essential resource for project team members who are looking to improve their project management skills. Whether you are new to project management or an experienced project manager, this book provides valuable insights and practical tips for breaking down projects into manageable tasks and sub-tasks. This book is a must-read if you want to improve your project management skills.

CHAPTER 2

Understanding Work Breakdown Structures

What is a WBS?

WBS stands for Work Breakdown Structure, a project management tool that helps break down a project into smaller, more manageable pieces. A WBS is a hierarchical representation of the tasks, activities, and deliverables that must be completed to achieve the project's objectives. It is an essential tool for project team members, as it helps ensure that all team members clearly understand the project's scope, timeline, and budget.

A WBS can be used for any type of project, regardless of size or complexity. It is especially useful for complex projects, as it allows team members to focus on specific tasks and deliverables rather than get overwhelmed by the project's overall complexity.

The WBS is typically created at the beginning of a project and is updated throughout the project's lifecycle. It is a living document that evolves as the project progresses and new information becomes available. The WBS is also used as a reference for project team members to ensure that they are all working towards the same goals and objectives.

A WBS is typically organized into a hierarchical structure, with the highest level representing the overall project and each subsequent level representing smaller, more specific tasks. The WBS can be broken down into as many levels as necessary to accurately represent the project's scope and deliverables.

In addition to helping to break down a project into smaller pieces, a WBS also helps to identify dependencies between tasks and deliverables. By identifying these dependencies, project team members can better understand how their work impacts the project as a whole and can work more effectively with other team members to ensure that all project goals are met.

Overall, a WBS is an essential tool for project team members. It helps ensure that all team members clearly understand the project's scope, timeline, and budget, allowing team members to focus on specific tasks and deliverables. By using a WBS, project team members can work more effectively together to ensure that the project is completed on time, within budget, and to the satisfaction of all stakeholders.

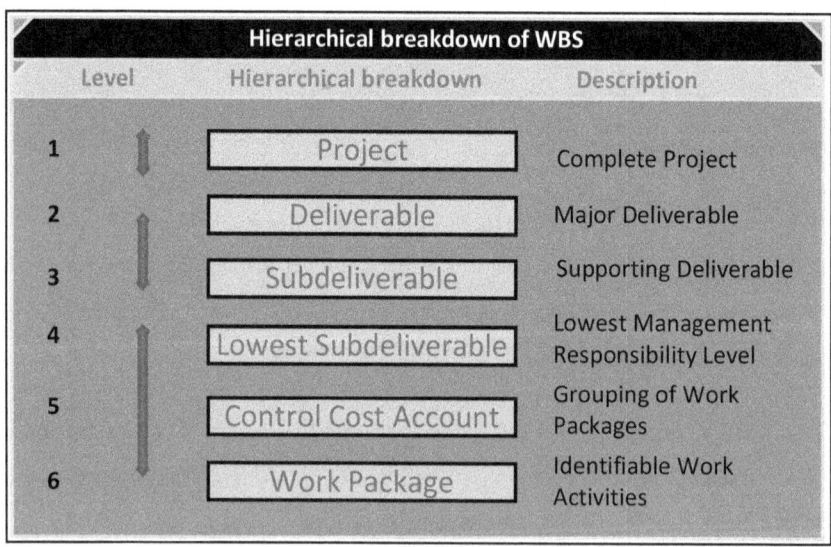

Figure 1 - Hierarchical Breakdown Structure

Why is WBS important in project management?

The Work Breakdown Structure (WBS) is a crucial tool in project management as it provides a comprehensive and structured approach to breaking down complex projects into smaller, manageable tasks. An effective WBS is essential for project success as it helps to define the scope of work, organize project tasks, and allocate resources efficiently.

One of the key benefits of a WBS is that it provides a clear understanding of the project deliverables and objectives. By breaking down the project into smaller, more manageable tasks, project team members can easily identify what needs to be accomplished and what tasks are dependent on others. This helps reduce ambiguity and ensures everyone works towards the same end goal.

Another benefit of a WBS is that it enables effective project planning and scheduling. By breaking down the project into smaller tasks, project managers can estimate the time and resources required for each task, and develop a realistic project schedule. This helps to ensure that the project is completed on time and within budget.

A well-designed WBS also provides a clear framework for project monitoring and control. By tracking progress against each task, project managers can identify potential issues early and take corrective action. This helps to reduce the risk of project delays and cost overruns.

Finally, a WBS provides a valuable communication tool for project team members, stakeholders, and sponsors. By providing a clear and structured overview of the project, everyone involved can easily understand the project scope, objectives, and progress. This helps to ensure that everyone is aligned and working towards the same end goal.

In conclusion, the WBS is a critical tool for project success. Breaking down complex projects into smaller, manageable tasks helps define the scope of work, organize project tasks, allocate resources efficiently, and ensure effective project monitoring and control. As a project team member, understanding the importance of the WBS and how to use it effectively is essential for project success.

Benefits of creating a WBS

Creating a Work Breakdown Structure (WBS) is essential in project management. This process involves breaking down a project into smaller, more manageable components, allowing for better planning, allocation of resources, and control over the project's progress. Here are some of the benefits of creating a WBS:

1. Helps in Better Planning

A WBS provides a clear and concise overview of the project, its objectives, and the scope of work. It helps identify the critical path, which is the sequence of activities that must be completed on time for the project to be successful. A WBS also helps identify dependencies between tasks, which is crucial for scheduling and resource allocation.

2. Facilitates Resource Allocation

A WBS helps identify the resources required for each component of the project. By breaking down the project into smaller components, it is easier to allocate resources based on the priority of each task. This makes resource allocation more efficient, as resources can be allocated based on the critical path.

3. Provides Better Control

A WBS provides a hierarchical structure of the project, allowing for better control over the project's progress. By breaking down the project into smaller components, it is easier to monitor the progress of each task and identify any delays or issues that may arise. This enables the project team to take corrective action in a timely manner, reducing the risk of delays and overruns.

4. Improves Communication

A WBS provides a common language for communicating project information. It helps align the project team's understanding of the project's scope, objectives, and deliverables. This improves communication between team members, stakeholders, and management, ensuring that everyone is on the same page.

5. Enhances Risk Management

A WBS helps identify potential risks and issues early on in the project. By breaking down the project into smaller components, it is easier to identify and mitigate risks at the component level, reducing the overall risk to the project.

In conclusion, a WBS is a valuable tool for project management. It helps in better planning, resource allocation, control, communication, and risk management. By creating a WBS, project team members can ensure that the project is completed successfully, on time, and within budget.

Types of WBS structures

A WBS is an essential project management tool that helps to break down complex projects into manageable tasks. This structure helps project managers define the work scope, estimate the effort required, allocate resources, and monitor progress. Project teams can use several types of WBS structures to achieve these goals.

We will discuss various types of WBS structures and give you a run-down of what they are and when one should use them. I have broken the WBS structures into three distinct sections. In Part I we explain and demonstrate the various WBS structures, in Part II, we will discuss the work breakdown structure dictionary. Finally, in part III, we show you what steps to follow to develop a proper WBS, with a sample guide for a project office.

Basic types of Structures: We will touch on the basics of the various structures that exist in business and within the project environment. As is explained quite clearly, a WBS always has a deliverable attached to it, secondly, it can be based on a specific timeline (phase or stage) of a project life-cycle. Lastly, we must set up an organization to drive the project to completion. Here are the basic WBS structures:

- Deliverable-based WBS
- Phase-based WBS
- Organizational WBS

Cardinal or Key Project Structures: Implementing these structures will assist the various disciplines to manage their deliverables with a clear objective. It must be noted that not all projects require all these WBS structures:

- Plant or Product? Pick one.
- Package - always required, most of the time orders must be placed.
- Systems - Big or small project, it is part of the handover process.
- Estimating - To get a detailed estimate for your project.
- Schedule - Can assist with a lower level of detail in the WBS.
- Cost – Manage your budget with control accounts.

Other Project Structures: Using these structures can guide disciplines with better management of their deliverables. These WBSs are very useful in large or mega projects:

- Resources - Works well in large global organizations.
- Risk - Integrate your risks into corporate Risks.
- Quality - The project requires very carefully managed quality.
- Stakeholder – For complicated and large/mega projects.
- Communication – Used in large organizations.

WBS Techniques: There are various methods to breakdown projects or business decisions, and we cover a few here that will assist. It is best to use these methods to get an outcome that will assist your organization:

- Responsibility - Allocates accountability and inputs.
- Network diagram- Assist teams in delving into the WBS details.
- Gantt charts- get a visual of your project timeline.
- CPA – Develop and create a Critical path.
- Mind Mapping - Assists in fleshing out complex issues.
- Affinity diagrams - Group business issues and decisions.
- Fishbone Diagrams - Used for cause-and-effect investigations.
- Decision Trees – Helps to make decisions.

In conclusion, project teams can choose from a variety of WBS structures depending on the nature of the project and the project management approach. Each has advantages and disadvantages. By choosing the right WBS structure, project teams can break down complex projects into manageable tasks, allocate resources effectively, and achieve project success.

Different Structures Across Project Life Cycles

Before we step off the different types of WBS structures, we have predictive, iterative, incremental, and agile life cycles.

Predictive WBS Structure:

In predictive project life cycles, often associated with the Waterfall methodology, the Work Breakdown Structure (WBS) is highly detailed and fixed from the start. The entire scope of the project, including all deliverables and activities, is defined at the beginning and broken down into manageable parts. This type of WBS is linear and sequential, reflecting the nature of the project's life cycle where each phase predictably follows the previous one without overlap.

Iterative WBS Structure:

Iterative life cycles involve a process where the project's scope is developed through repeated cycles. The WBS in an iterative life cycle is more flexible and revisited throughout the project's duration. It allows for more frequent reassessment and adaptation, which is reflected in the WBS's ability to incorporate changes and new information as the project progresses through its iterations.

Incremental WBS Structure:

In incremental life cycles, the project is delivered in parts, with each increment providing a portion of the functionality. The WBS for incremental projects is divided into smaller, more manageable components that can be completed in successive phases. This approach allows for portions of the project to be delivered and potentially used before the entire project is complete, facilitating feedback and continuous improvement.

Agile WBS Structure:

Agile project life cycles are characterized by their flexibility, adaptability, and customer-centric approach. The WBS within an Agile project is often less rigid and more high-level compared to predictive life cycles. It focuses on delivering small, workable features in short cycles called sprints. The WBS is continuously updated to reflect changes in the project backlog, with each sprint delivering a potentially shippable product increment.

Each of these WBS structures is tailored to the corresponding project life cycle, ensuring that the breakdown of work aligns with the overall approach to project management. Whether through a predictive, iterative, incremental, or agile life cycle, the WBS remains a fundamental tool for project managers to plan, execute, and monitor project progress effectively.

Deliverable-based WBS

A deliverable-based work breakdown structure (WBS) is a hierarchical decomposition of the project scope into smaller, more manageable components known as deliverables. These deliverables are concrete, measurable outputs that the project team must produce to complete the project.

The primary advantage of a deliverable-based WBS is that it helps project team members focus on what needs to be delivered and not just what activities need to be performed. By breaking down the project scope into deliverables, the project team can identify the specific outputs that need to be produced and then plan and execute the required activities.

To develop a deliverable-based WBS, the project team should follow these steps:

1. Identify the project objectives: The project team should identify the high-level goals and objectives of the project. These objectives should be specific, measurable, achievable, relevant, and time-bound.

2. Define the project scope: The project team should define the project's boundaries, including the in-scope and out-of-scope items.

3. Identify the deliverables: The project team should identify the specific outputs that the project must produce to achieve the project objectives. These deliverables should be tangible, measurable, and verifiable.

4. Organize the deliverables: The project team should organize the deliverables into a hierarchical structure. The top-level deliverables represent the major project outputs, while the lower-level deliverables represent the sub-components required to produce the top-level deliverables.

5. Assign responsibilities: The project team should assign responsibilities for producing each deliverable to specific team members. This helps to

15

ensure that everyone understands their roles and responsibilities and can work together to deliver the project.

6. Estimate effort: The project team should estimate the effort required to produce each deliverable, including the activities required to produce the deliverables.

7. Develop the project schedule: The project team should use the deliverable-based WBS to develop the project schedule. The schedule should identify each deliverable's start and end dates and the dependencies between the deliverables.

In conclusion, a deliverable-based WBS is an effective tool for breaking down the project scope into smaller, more manageable components. By focusing on the specific outputs that the project must produce, the project team can plan and execute the activities required to deliver the project successfully. The deliverable-based WBS helps to ensure that everyone understands their roles and responsibilities and can work together to achieve the project objectives.

A WBS is a critical non-negotiable tool in project management, particularly in disciplines like Project Services and Controls. It helps in breaking down the project into smaller, more manageable pieces. Below is a typical sample WBS for an engineering project, structured in a WBS and table format. This WBS is designed to be comprehensive and has 3 levels to ensure that all aspects of the engineering are covered for all required disciplines.

Here is a typical deliverable-based work breakdown structure.

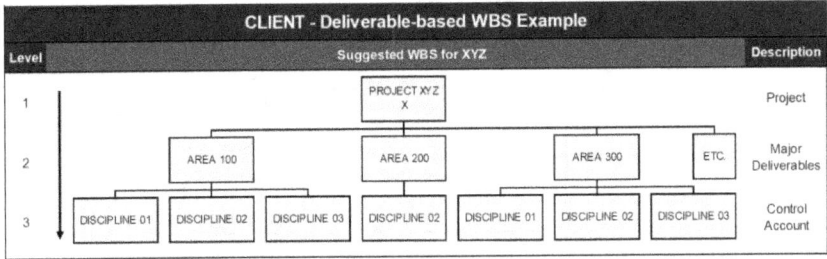

Figure 2 - Sample -Deliverable-based WBS

WBS Coding				
Level	WBS Number	#	Description	Detail
1	X	X	Project XYZ	Project XYZ Description
2	X.100	100	Area 100	Area 100 Description
3	X.100.01	01	Discipline 01	Discipline 01 Description
3	X.100.02	02	Discipline 02	Discipline 02 Description
3	X.100.03	03	Discipline 03	Discipline 03 Description
2	X.200	200	Area 200	Area 200 Description
3	X.200.01	01	Discipline 01	Discipline 01 Description
2	X.300	300	Area 300	Area 300 Description
3	X.300.01	01	Discipline 01	Discipline 01 Description
3	X.300.02	02	Discipline 02	Discipline 02 Description
3	X.300.03	03	Discipline 03	Discipline 03 Description
	Etc.			

Table 1 - Deliverable-based WBS Coding

This WBS is aligned with standard project management methodologies and can serve as a robust framework for managing the complexities involved in building a house. It allows for better control and monitoring, ensuring the project stays on schedule and within budget.

Level 1	Level 2	Level 3
Construction of Guest House	Project Management	Initiation
		Planning
		Execution
		Monitoring and Control
		Closure
	Site Preparation	Site Selection
		Clearing and Grading
		Soil Testing
	Foundation	Excavation
		Footings
		Foundation Walls
		Slab Pouring
	Structural Framing	Framing Design
		Framing Material Procurement
		Wall Framing
		Roof Framing
	Exterior Finishing	Siding Installation
		Roofing
		Windows and Doors
		Exterior Painting
	Interior Finishing	Electrical Wiring
		Plumbing Installation
		Drywall Installation
		Flooring
		Interior Painting
	Utilities	Water Supply
		Sewer System
		Gas Connection
		Electrical Connection
	Landscaping	Garden Planting
		Driveway Construction
	Quality Assurance	Inspections & Testing
		Quality Control Measures
	Regulatory Compliance	Permit Acquisition
		Compliance Documentation

Table 2 - A typical House Construction project WBS

Phase-based WBS

A phase-based WBS is a work breakdown structure that is organized around project phases or stages. It is a useful tool for project team members because it provides a clear picture of what work needs to be done in each project phase. By breaking the project down into smaller, more manageable pieces, the team can more easily track progress, identify potential issues, and make adjustments as needed.

To create a phase-based WBS, start by identifying the major phases or stages of the project. These could include planning, design, implementation, testing, and maintenance. Each phase should be broken down into smaller tasks or activities, which are then further broken down into sub-tasks or work packages. The result is a hierarchical structure that shows the relationship between the different levels of work.

One of the benefits of a phase-based WBS is that it allows project team members to focus on one phase at a time rather than trying to tackle the entire project all at once. This can help to reduce stress and increase productivity, as team members can concentrate on completing one phase before moving on to the next.

Another advantage of a phase-based WBS is that it can help to identify potential issues early on in the project. Breaking the project down into smaller pieces makes it easier to see where problems might occur and take corrective action before they become larger issues.

Finally, a phase-based WBS can help ensure that all necessary tasks are completed on time. Breaking the project down into smaller tasks and assigning each task to a specific phase makes it easier to track progress and ensure that everything is on track.

In conclusion, a phase-based WBS is useful for project team members because it provides a clear roadmap for completing the project. By breaking the project down into smaller, more manageable pieces, the project team can stay focused, identify potential issues, and ensure that all necessary tasks are completed in a timely manner.

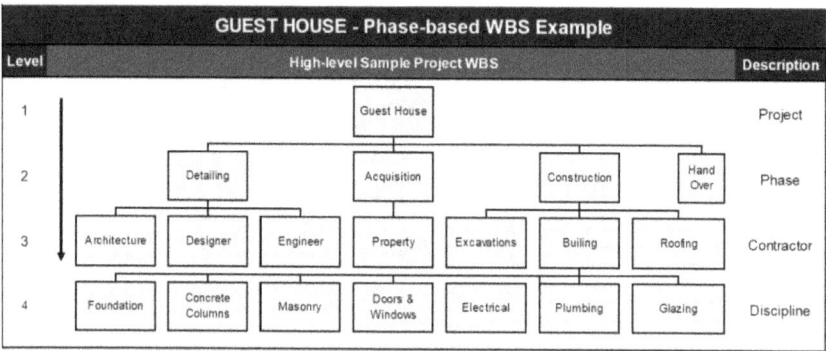

Figure 3 - Simple Guest House - Phase-based WBS

Organizational-based WBS

Organizational-based WBS is a type of work breakdown structure that is organized according to the departments or functions within an organization. It is often used in large organizations with multiple departments or business units working on a project. The organizational-based WBS is structured to align with the organization's structure and processes, making it easier to manage and track progress.

In an organizational-based WBS, the project is divided into smaller components based on the different departments or functions involved in the project. Each component is then broken down further into smaller tasks that must be completed. This allows for better communication and coordination between different departments, ensuring everyone is on the same page and working towards the same objectives.

One of the advantages of using an organizational-based WBS is that it allows for better resource allocation. Since each department is responsible for a specific part of the project, resources can be allocated more efficiently. For

20

example, suppose the project involves both the marketing and engineering departments. In that case, the marketing department can focus on the marketing aspects of the project while the engineering department focuses on the technical aspects.

Another advantage of an organizational-based WBS is that it helps to identify potential issues early on in the project. Since each department is responsible for a specific part of the project, any issues or problems can be identified and addressed before they become bigger problems. This helps to ensure that the project stays on track and within budget.

However, there are also some disadvantages to using an organizational-based WBS. One of the main disadvantages is that it can lead to silos within the organization. Each department may focus only on its specific part of the project without considering how it fits into the bigger picture. This can lead to a lack of collaboration and communication between departments.

An organizational-based WBS can be a useful tool for managing large projects in complex organizations. It allows for better resource allocation, improved communication and coordination, and early identification of potential issues. However, it is important to ensure that all departments are working towards the same objectives and that there is open communication and collaboration between departments to avoid silos.

Below is a typical Small to Medium Engineering House Organizational Breakdown Structure (OBS). As is clearly demonstrated in the OBS below, it allows the organizational hierarchy to be arranged in organizational units or departments. Inside each department you will find various disciplines.

The vacant positions are shown in the yellow boxes, allowing for organizational growth in the future.

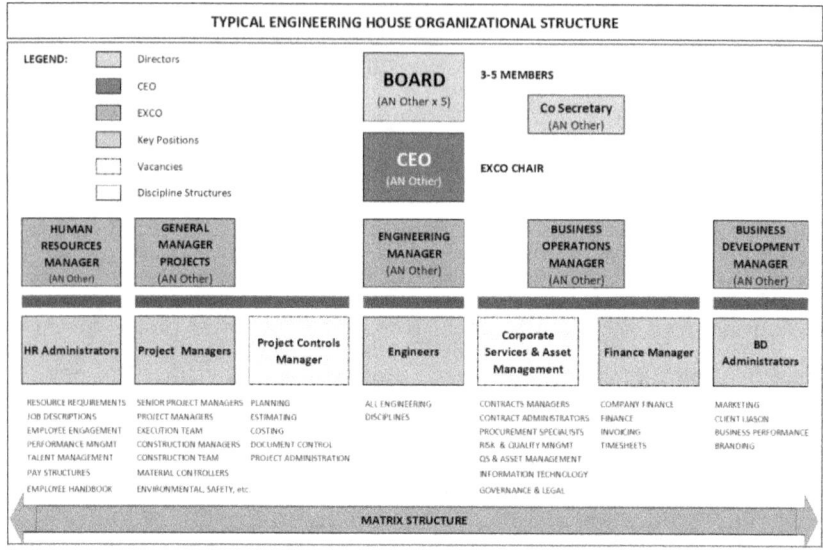

Figure 4 - Sample – Organizational Breakdown Structure (OBS)

Creating a WBS

The Work Breakdown Structure (WBS) is a crucial component of project management that helps teams break down complex projects into smaller, more manageable tasks. It provides a systematic approach to project planning, scheduling, and resource allocation, ensuring that every aspect of the project is accounted for and completed on time.

Creating a WBS involves breaking down the project into smaller, more manageable pieces, known as work packages. These work packages should be clearly defined and assigned to specific team members or groups, with individual tasks and milestones identified.

To create a WBS, the project team should follow a few simple steps:

1. Identify the project objectives and deliverables: The first step in creating a WBS is to identify the project objectives and deliverables. These should be specific, measurable, achievable, relevant, and time-bound, and should be agreed upon by all stakeholders.

2. Break down the project into phases or stages: Once the project objectives and deliverables have been identified, the project team should break down the project into smaller, more manageable phases or stages. These phases or stages should be based on logical groupings of tasks and should be defined in terms of their objectives and deliverables.

3. Identify the work packages: The project team should identify the work packages within each phase or stage. These are the smallest units of work that can be assigned to a specific team member or group and should be clearly defined in terms of their objectives, deliverables, and timelines.

4. Assign tasks and milestones: Once the work packages have been identified, the project team should assign tasks and milestones to each work package. These tasks should be clearly defined and should be assigned to specific team members or groups, with individual deadlines and milestones clearly identified.

5. Review and refine the WBS: Finally, the project team should review and refine the WBS, ensuring that all tasks and milestones are clearly defined and assigned and that the WBS accurately reflects the objectives and deliverables of the project.

Creating a WBS is a critical component of project management and can help teams manage complex projects more effectively. By breaking down the project into smaller, more manageable pieces, the project team can ensure that every aspect of the project is accounted for and completed on time.

Steps involved in creating a WBS

Creating a Work Breakdown Structure (WBS) is a critical step in project planning. It involves breaking down the project into smaller, more manageable pieces that can be assigned to different team members and tracked for progress. A well-designed WBS provides a clear roadmap for the project and ensures everyone is on the same page.

Here are the steps involved in creating a WBS:

Step 1: Define the Project Scope

Before you can start breaking down the project, you need to define its scope. This includes identifying the objectives, deliverables, and constraints of the project. You should also consider any risks or assumptions that might impact the project.

Step 2: Identify Major Deliverables

Next, you should identify the major deliverables of the project. These high-level items need to be completed to achieve the project objectives. You can use brainstorming or mind mapping techniques to identify these deliverables.

Step 3: Divide Deliverables into Sub-Deliverables

Once you have identified the major deliverables, you can start breaking them down into sub-deliverables. This involves dividing each major deliverable into smaller chunks that individual team members can complete. You should aim to make each sub-deliverable as specific as possible.

Step 4: Assign Responsibility

After you have broken down the deliverables into sub-deliverables, you should assign responsibility for each one to specific team members. This ensures that everyone knows what they are responsible for and can focus on their specific tasks.

Step 5: Create a WBS Diagram

The final step is to create a WBS diagram. This is a visual representation of the project and its deliverables. The WBS diagram should show the hierarchy of deliverables, sub-deliverables, tasks, and the responsible team members.

In conclusion, creating a WBS is an essential step in project planning. By following these steps, you can ensure that your project is broken down into manageable pieces and everyone knows their responsibility. This will help you to stay on track and achieve your project objectives.

Common mistakes to avoid

As a project team member, you play a critical role in the success of your project. While your focus may be on executing your assigned tasks, it's important to be aware of common mistakes that can derail a project. Here are some of the most common mistakes to avoid:

1. Lack of communication: Communication is key in any project. Make sure you are communicating regularly with your team members and stakeholders. Be clear and concise in your communication, and don't be afraid to ask questions or seek clarification. A lack of communication can lead to misunderstandings, missed deadlines, and a breakdown in team morale.

2. Lack of Understanding: One of the common challenges in implementing a WBS is a lack of understanding of its purpose and benefits. Project teams may not fully comprehend how a WBS can help organize project tasks, identify dependencies, and allocate resources. To address this challenge, it is essential to provide comprehensive training and education to the project team members about the importance and benefits of using a WBS.

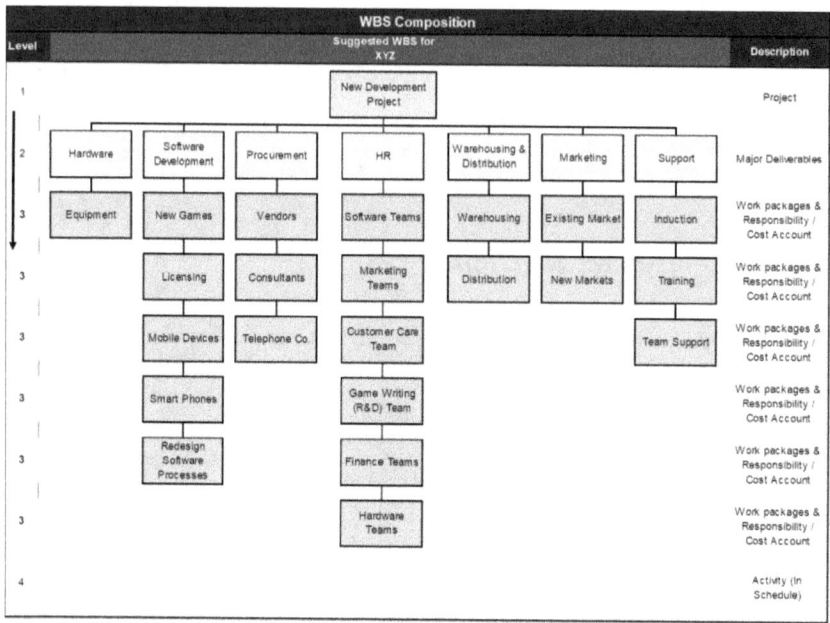

Figure 5 - New Software development WBS

WBS No	LEVEL 1	LEVEL 2	LEVEL 3	DESCRIPTION	RESOURCE	DETAIL	PACKAGE
N	N			New Development Project	PM	Deliver new software and hardware development	
N1	N	1		Hardware	HW	Handover hardware for project	
N2	N	2		Software Development	SW	Handover software for project	
N3	N	3		Procurement	PR	Procure all packages	
N4	N	4		HR	HR	Support all HR functions and appointments	
N5	N	5		Warehousing & Distribution	WD	Ensure that Warehousing & Distribution is adequate	
N6	N	6		Marketing	MR	Supply powerful marketing for products	
N7	N	7		Support	SP	Support function of the products	
N1E	N	1	E	Equipment	HW	Purchase new equipment	N1-01
N2N	N	2	N	New Games	SW	Develop new games	N2-01
N2L	N	2	L	Licensing	SW	Ensure that licensing conforms to legal requirements	N2-02
N2M	N	2	M	Mobile Devices	SW	Develop new games for mobile devices	N2-03
N2S	N	2	S	Smart Phones	SW	Develop new games for smart devices	N2-04
N2R	N	2	R	Redesign Software Processes	SW	Redesign existing software processes	N2-05
N3V	N	3	V	Vendors	PR	Appoint vendors	
N3C	N	3	C	Consultants	PR	Appoint consultants	
N3T	N	3	T	Telephone Co.	PR	Negotiate with Telephone company	
N4S	N	4	S	Software Teams	HR	Develop Software Teams	
N4M	N	4	M	Marketing Teams	HR	Develop Marketing Teams	
N4C	N	4	C	Customer Care Team	HR	Develop Customer Care Team	
N4G	N	4	G	Game Writing (R&D) Team	HR	Develop Game Writing (R&D) Team	
N4F	N	4	F	Finance Teams	HR	Develop Finance Teams	
N4H	N	4	H	Hardware Teams	HR	Develop Hardware Teams	
N5W	N	5	W	Warehousing	WD	Expand required warehousing	
N5T	N	5	T	Distribution	WD	Expand required distribution	
N6E	N	6	E	Existing Market	MR	Review existing markets	
N6N	N	6	N	New Markets	MR	Develop and investigate new markets	
N7D	N	7	D	Induction	SP	Develop an induction process	N7-01
N7R	N	7	R	Training	SP	Develop a internal training process	N7-02
N7S	N	7	S	Team Support	SP	Appoint a support team for the new developments	N7-03

Table 3 - New Software Development WBS Coding

3. Poor planning: Planning is essential for any successful project. Make sure you are taking the time to plan out your tasks and deadlines and considering any potential roadblocks or risks. Failing to plan can lead to missed deadlines, budget overruns, and a lack of direction for the project.

4. Scope creep: Scope creep is when the project scope expands beyond what was originally planned. This can happen when new requirements are added to the project without proper evaluation of the impact on the timeline and budget. Ensure you are closely monitoring the project scope and that any changes are properly evaluated and approved.

5. Lack of accountability: Each team member should be accountable for their assigned tasks. Make sure you are tracking your progress and reporting any issues or delays to your project manager. Failing to take responsibility for your tasks can lead to missed deadlines and a breakdown in team morale.

6. Ignoring risk management: Every project has risks. It's important to identify potential risks and develop a plan to mitigate them. Failing to address risks can lead to costly delays and unexpected problems.

7. Resistance to Change: Resistance to change is a common challenge faced by project teams when implementing any new project management tool, including a WBS. Some team members may be resistant to adopting a WBS due to unfamiliarity or concerns about additional workload. To address this challenge, project leaders should provide adequate training, support, and emphasize the benefits of using a WBS in project planning and execution.

8. Lack of Standardization: Inconsistent use of terminology and structure can undermine the effectiveness of a WBS. Lack of standardization may lead to confusion and miscommunication among team members. It is crucial for project teams to establish a standardized WBS dictionary that defines the terminology and structure to ensure consistency and clarity throughout the project. I have added a complete section (a few chapters) where we discuss the Work Breakdown Structure Dictionary (WBSD) in detail.

A WBSD is a very important document to keep track of during the life cycle of a project. The biggest mistake projects make is not keeping the WBSD updated during the execution, monitoring & control, and close-out phases of a project. Project teams normally only focus on the dictionary during the planning phase of a project and neglect to update the WBSD during the later phases. For example, the WBSD (besides the baseline schedule) becomes one of the most valuable items of information during a contractual dispute when it comes to an extension of time (EOT). Any changes that occur, should be logged to make any dispute easier to manage and close out.

Implementing a proper WBS can significantly enhance project planning and execution. However, project teams may encounter challenges such as a lack of understanding, scope creep, communication issues, resistance to change, and lack of standardization. By recognizing these challenges and implementing appropriate strategies, project teams can overcome these obstacles and successfully implement a WBS to improve project management outcomes.

By avoiding these common mistakes, you can help ensure the success of your project. Remember, as a project team member, you have a critical role to play in the project's success. Take ownership of your tasks, communicate regularly with your team members and stakeholders, and be proactive in identifying and addressing potential issues.

Key Tips: Understanding a WBS

As I've learned about Work Breakdown Structure (WBS), I've found a few key tips that have helped me understand and effectively use this project management tool.

Start with a clear project objective: Before diving into creating a WBS, it's important to have a clear understanding of the project objective. This will help you break down the work in a logical and organized manner.

Break down the project into manageable chunks: The purpose of a WBS is to break down the project into smaller, more manageable tasks. It's crucial to break down the work into meaningful components that can be easily understood and executed.

Use a hierarchical structure: A WBS is typically presented in a hierarchical format, where the project is divided into major deliverables or phases, which are further divided into sub-deliverables or tasks. This hierarchical structure helps visualize the project's scope and ensures no work is overlooked. I have included various structures in this book to demonstrate the WBS and other structures to help you manage your project(s) more effectively.

Involve the project team: Creating a WBS should not be a one-person task. It's important to involve the project team, as they have valuable insights and expertise that can contribute to the breakdown of the work. Collaborating with the team also ensures buy-in and accountability from all members.

Use consistent and concise naming conventions: When creating the WBS, it's essential to use consistent and concise names for each task or deliverable. This helps to avoid confusion and ensures clarity throughout the project.

Consider dependencies and relationships: While breaking down the work, it's crucial to identify any dependencies or relationships between tasks. This helps understand the sequence of activities and ensures the project flows smoothly.

Keep it flexible: A WBS is not set in stone and can be adjusted as the project progresses. It's important to regularly review and update the WBS and the WBSD to accommodate any changes or new information that may arise during the project lifecycle.

Use software tools: Various project management software tools are available to help create and manage a WBS effectively. Utilizing these tools can enhance productivity and provide a centralized platform for collaboration and tracking.

By following these key tips, I've been able to better understand and utilize the Work Breakdown Structure in my project management endeavours. It has helped me to break down complex projects into manageable tasks, ensure clarity and accountability, and ultimately achieve successful project outcomes.

Lessons Learnt: Understanding a WBS

As I reflect on my experience with understanding Work Breakdown Structure (WBS), I have learned several valuable lessons.

Firstly, I have come to appreciate the importance of having a clear and well-defined WBS in any project or task. It serves as a roadmap, providing a comprehensive breakdown of the work that needs to be done and the deliverables that need to be achieved. Having a clear WBS makes it easier to understand the project's scope and ensure that all necessary tasks are accounted for.

Secondly, I have learned that creating a WBS requires a careful analysis and understanding of the project requirements. It is crucial to have a deep understanding of the project objectives and the different components and tasks involved. This understanding allows for the creation of a detailed and accurate WBS that aligns with the project's goals.

Additionally, I have realized the importance of involving key stakeholders in creating and reviewing the WBS. Different perspectives and insights can be gathered by consulting with team members and other relevant parties, leading to a more comprehensive and effective WBS. It also ensures that everyone is on the same page and clearly understands the project's structure and timeline.

Furthermore, I have learned that the WBS should be flexible and adaptable. As projects progress, changes or unforeseen circumstances may require adjustments to the original WBS. It is essential to be open to modifications and updates, ensuring that the WBS remains relevant and reflective of the project's needs.

Lastly, I have discovered the importance of regularly reviewing and updating the WBS throughout the project lifecycle. By continuously monitoring and evaluating the progress of tasks and deliverables, any deviations or issues can be identified and addressed promptly. This allows for better control and management of the project, ensuring it stays on track and meets its objectives.

In conclusion, understanding the Work Breakdown Structure (WBS) has taught me the significance of having a clear and well-defined roadmap for any project or task. It has emphasized the need for careful analysis, stakeholder involvement, flexibility, and regular review and updates. By incorporating these lessons, you should feel confident in your ability to effectively utilize and understand the WBS's role in future projects.

NOTE: Make sure that you do not just list or group reporting titles as your WBS. Using MS Projects restricts you from using an elaborate coding system that allows you to review your project reports in various areas. I.e. Sort by Area, Discipline. Then sort by: Discipline, Package No, and Responsibility or Discipline, Responsibility, and Package No.

Some tools assist you in developing a proper report that makes sense to everyone and helps you guide or direct your team to work on the right deliverables.

CHAPTER 3

Cardinal or Key Project Structures

What does Cardinal mean?

Besides the first three WBS structures discussed in the previous chapter, these cardinal or prime or key breakdown structures are required to produce a definitive project scope. These are the recommended structures you should use to ensure you deliver a successful project. In some cases, the original WBS will encompass all these structures' information, and we do cover that in depth in the WBS dictionary section. It must be noted that the bigger and more complicated a project is, the more you should break the project down and allocate the responsibilities to various structures for team members to manage their work down to lower levels. These structures are generally embedded during the execution phase of my projects.

NOTE: I am not going to discuss Area breakdown structures (ABS) as they are normally encompassed into the various WBS elements at quite a high level, and resources are allocated in the breakdown structure as area managers. For bigger projects, I recommend implementing a KPI and KRA system that will keep the resources focused on their areas of accountability, which in turn will drive for a better successful delivery. However, when risk assessments are performed during the various project phases, all Areas should form part of that risk assessment, as there might be risks in some areas that will impact

*other areas. Mitigation steps could be implemented,
and the most appropriate person should be allocated to
manage those risks.*

*Depending on your project, you might want to use a
product breakdown structure instead of a plant
breakdown structure. This comes in handy when you
repeat the project over and over, of which
manufacturing is a good example. There might be
various opinions on this, but feel free to use what fits
best for your deliverables. Please take note of the
deliverable-based WBS, which I will cover with the
following two structures.*

Plant Breakdown Structure (PBS)

The Plant Breakdown Structure (PBS) is a hierarchical representation of a plant or facility's components, systems, and subsystems. It provides a detailed breakdown of the various elements that comprise the plant, allowing for a systematic analysis and understanding of its structure.

At the top level of the PBS, the plant is divided into major sections, such as the main building, processing units, utilities, storage areas, and support systems. Each section is then further broken down into smaller components, which may include equipment, machinery, piping, electrical systems, instrumentation, and control systems.

The PBS provides a comprehensive plant overview, enabling stakeholders to visualize and comprehend its complexity. It helps identify the interdependencies between different elements and ensures that all aspects of the plant are considered during the design, construction, operation, and

34

maintenance phases.

Level	Task	Level	Task Continued
1	**Construction of Guest House**	2	**Exterior Finishing**
2	**Project Management**	3	Siding Installation
3	Initiation	3	Roofing
3	Planning	3	Windows and Doors
3	Execution	3	Exterior Painting
3	Monitoring and Control	2	**Interior Finishing**
3	Closure	3	Electrical Wiring
2	**Site Preparation**	3	Plumbing Installation
3	Site Selection	3	Drywall Installation
3	Clearing and Grading	3	Flooring
3	Soil Testing	3	Interior Painting
2	**Foundation**	2	**Utilities**
3	Excavation	3	Water Supply
3	Footings	3	Sewer System
3	Foundation Walls	3	Gas Connection
3	Slab Pouring	3	Electrical Connection
2	**Structural Framing**	2	**Landscaping**
3	Framing Design	3	Lawn Installation
3	Framing Materials Procurement	3	Garden Planting
3	Wall Framing	3	Driveway Construction
3	Roof Framing	2	**Quality Assurance**
2	**Permitting & Regulatory**	3	Inspections
3	Permit Acquisition	3	Quality Control Measures
3	Compliance Documentation	3	Testing

Table 4 - Guest House WBS Levels – Outline Style

The table above reflects a typical detailed work breakdown structure (WBS) that could be considered for a guest house project.

The hierarchical nature of the PBS allows for a clear understanding of the relationships between the various components. For instance, a piece of equipment may be part of a larger system, which in turn is part of a section. This structure helps organise and manage the plant effectively, allowing for easy navigation and efficient communication among different teams involved in plant operations.

Moreover, the PBS is a foundation for other project management activities, such as cost estimation, scheduling, risk assessment, and resource allocation.

35

It provides a framework for breaking down the work scope and facilitates the identification of deliverables, milestones, and dependencies. A PBS can be huge (depending on the product) and help ensure that all necessary activities are accounted for and that the project progresses smoothly.

The Plant Breakdown Structure (PBS) is a crucial tool in plant management and project execution. It provides a detailed breakdown of the plant's structure, allowing for a systematic analysis of its components and systems. By organizing and visualizing the plant in a hierarchical manner, the PBS facilitates effective communication, coordination, and decision-making throughout the plant's lifecycle.

Project management is crucial in managing the construction, renovation, or expansion of various types of plants and facilities. Here are some examples of plants or facilities that can be managed effectively using WBS principles:

- **Manufacturing Plant:** Managing the construction or renovation of a manufacturing facility for producing goods, including equipment installation, process design, and quality control systems.
- **Power Plant:** Overseeing the construction of power generation facilities such as coal, nuclear, natural gas, or renewable energy plants.
- **Chemical Processing Plant:** Managing projects related to chemical manufacturing facilities, including safety protocols, environmental compliance, and process optimization.
- **Water Treatment Plant:** Coordinating projects to upgrade or expand water treatment facilities to meet increased demand or stricter regulatory requirements.
- **Wastewater Treatment Plant:** Managing projects to improve or expand wastewater treatment facilities for more efficient and environmentally friendly operations.
- **Oil Refinery:** Overseeing projects for the construction,

maintenance, or upgrades of oil refineries, including process optimization and safety measures.

- **Food Processing Facility:** Planning and executing projects to build or renovate food processing plants, ensuring compliance with food safety standards.

- **Pharmaceutical Manufacturing Facility:** Managing projects related to pharmaceutical production plants, including compliance with regulatory requirements.

- **Hospitals and Healthcare Facilities:** Coordinating construction or renovation projects for healthcare facilities, such as hospitals, clinics, or medical research centers.

- **Educational Institutions:** Managing construction projects for schools, colleges, and universities, including campus expansions, new building construction, and infrastructure upgrades.

- **Commercial Buildings:** Overseeing the construction or renovation of commercial properties, including office buildings, shopping centers, and hotels.

- **Sports Arenas and Stadiums:** Planning and managing the construction or renovation of sports and entertainment venues, including seating, facilities, and infrastructure.

- **Transportation Hubs:** Coordinating projects for transportation facilities like airports, ports, and railway stations, including terminal expansions and infrastructure improvements.

- **Data Centers:** Managing projects related to the construction or upgrade of data centers to support growing technology needs.

- **Municipal Facilities:** Overseeing projects for municipal buildings such as city halls, libraries, fire stations, and police stations.

- **Manufacturing Facilities:** Planning and managing the setup of new manufacturing facilities or optimizing existing ones.

- **Distribution Centers:** Coordinating the construction or expansion of warehouses and distribution centers to meet increased demand.

- **Iron Ore Processing Plant:** Overseeing the construction and operation of an iron ore processing plant, which includes the installation of crushers, conveyors, and processing equipment for ore beneficiation.

- **Natural Gas Processing Plant:** Managing a project to construct and operate a natural gas processing facility, involving gas extraction, purification, compression, and distribution.

- **Salt Evaporation Plant:** Coordinating the construction and operation of a salt evaporation plant, where seawater or brine is processed to produce high-quality salt products.

- **Phosphate Mining and Processing Plant:** Planning and executing a project for a phosphate mining and processing facility, including mining operations, beneficiation processes, and fertilizer production.

- **Uranium Enrichment Plant:** Overseeing the construction and operation of a uranium enrichment facility, ensuring compliance with nuclear safety regulations and security protocols.

- **Precious Metals Processing Facility:** Managing a project to build a processing plant for rare earth elements, which includes mineral extraction, chemical processing, and refining processes.

These examples demonstrate the diverse range of plants and facilities that benefit from practical project management to ensure successful planning, execution, and completion of projects within budget and on schedule.

Difference between Product and Plant

I often draw parallels between the **Product Breakdown Structure (PBS1)** (numbered here to support the explanation) and the **Plant Breakdown Structure (PBS2)**, both of which are subsets of the overarching Work

Breakdown Structure (WBS) methodology. While the PBS1 dissects a product into its constituent parts, the Plant Breakdown Structure does the same for a physical facility, detailing its systems and components. Both structures serve to crystallize the scope of work, yet they cater to different domains: the PBS1 is product-focused, ideal for tangible deliverables and often used in manufacturing or product development, whereas the Plant Breakdown Structure is facility-oriented, crucial in construction or plant setup projects.

The WBS, the umbrella under which both structures fall, is the common thread that ensures every element of a project, whether it be a product or a plant, is captured. The WBS facilitates a comprehensive view, breaking down a project into manageable tasks and deliverables. The key similarity lies in their hierarchical nature, which helps in identifying all necessary tasks and preventing scope creep. However, the distinction comes from their application; the PBS1 is pivotal for understanding how product components interact and are integrated, while the Plant Breakdown Structure is essential for visualizing the layout and interdependencies of plant systems.

Choosing between these structures depends on the project's end goal. For product development, the PBS1 is indispensable for its detailed focus on the end product. In contrast, for projects aiming to establish or modify plant facilities, the Plant Breakdown Structure is more appropriate, providing clarity on the spatial and systemic organization of the plant. Both, however, are integral to the WBS, ensuring that every aspect of the project is planned, accounted for, and aligned with the project's objectives.

Product Breakdown Structure (PBS)

The Product Breakdown Structure (PBS) is a hierarchical diagram or document that breaks down a product into its component parts or deliverables. It is a tool commonly used in project management to define and organize the scope of work required to deliver a product or system. The PBS provides a clear and structured representation of the product, allowing project teams to understand the various elements that need to be developed, designed, or procured.

At the top level of the PBS, the product is depicted as a single entity, which is then progressively decomposed into smaller and more manageable components. Each level of the PBS represents a different level of detail, with the lowest level being the individual deliverables or work packages that can be assigned to specific team members.

The PBS helps project teams to define and clarify the boundaries of the product being developed, ensuring that all elements are accounted for and nothing is overlooked. It provides a visual representation of the product's structure, allowing stakeholders to understand how the different parts fit together and how changes in one area may impact other components. This promotes effective communication, coordination, and collaboration among team members.

Furthermore, the PBS serves as a foundation for other project management activities, such as estimating costs and resources, creating a project schedule, and identifying dependencies between different deliverables. It helps identify the project's critical path, as well as potential risks and challenges that may arise during the development process.

One of the key benefits of using a PBS is that it enables project teams to clearly understand the scope of work, preventing scope creep and ensuring that all necessary tasks are accounted for. It also facilitates better resource allocation, as team members can easily identify their specific responsibilities and tasks.

In conclusion, the Product Breakdown Structure is a vital tool in project management that helps to define, organize, and communicate the scope of work required to deliver a product. It visually represents the product's structure and helps project teams plan, coordinate, and execute their work effectively. By breaking down the product into its component parts, the PBS ensures that all elements are accounted for, promotes effective communication, and facilitates better resource allocation.

Project management is a versatile discipline that can be applied to various products and projects across different industries. Here are just a few samples of products that can be managed using our WBS techniques:

- **Software Development:** Managing the creation of a new software application, from initial planning and requirements gathering to coding, testing, and deployment.

- **Construction Projects:** Overseeing the construction of a building, bridge, or infrastructure project, including scheduling, resource allocation, and quality control.

- **Product Launch:** Coordinating the launch of a new product, from market research and product development to marketing and sales strategies.

- **Event Planning:** Organizing and executing events like conferences, weddings, or trade shows, including logistics, budgeting, and scheduling.

- **Marketing Campaigns:** Managing marketing campaigns, including defining objectives, creating content, and tracking performance metrics.

- **Manufacturing:** Supervising the production process of goods, ensuring efficiency, quality control, and meeting production deadlines.

- **Research and Development:** Leading R&D projects for the development of new technologies, products, or innovations.

- **Healthcare Projects:** Implementing projects to improve healthcare services, such as hospital expansions, IT system upgrades, or patient care initiatives.

- **Infrastructure Projects:** Planning and managing large-scale infrastructure developments like highways, airports, and public transportation systems.

- **Film Production:** Coordinating all aspects of a movie or TV show production, from script development to post-production and distribution.

- **Website Development:** Managing the creation or redesign of a website, including content creation, design, coding, and testing.

- **Product Manufacturing:** Overseeing the manufacturing process of physical products, ensuring quality, cost control, and timely delivery.

- **Educational Programs:** Developing and managing educational programs or courses, including curriculum design, teacher training, and student assessment.

- **Nonprofit Initiatives:** Managing projects for nonprofit organizations, such as fundraising campaigns, community development projects, or disaster relief efforts.

- **Aerospace Projects:** Leading complex aerospace projects like aircraft design, manufacturing, and space exploration missions.

- **Coal Mine:** Developing and managing a project to establish a new coal mine, including site selection, environmental assessments, infrastructure construction, and the implementation of mining operations.

- **Gold Mine:** Planning and executing a project to establish a gold mining operation, which involves geological surveys, mine shaft construction, ore extraction, and refining processes.

- **Copper Mine:** Managing the construction and operation of a copper mining facility, including the design of processing plants, equipment procurement, and mining operations.

- **Diamond Mine:** Coordinating the development of a diamond mine, which encompasses geological exploration, diamond extraction methods, and the establishment of sorting and cutting facilities.

These are just a few examples of products and projects that benefit from project management practices. Project management provides a structured approach to planning, executing, and controlling a wide range of endeavours, making it an essential tool in various industries and sectors.

Task	Description	Subtasks
Project: Develop Software	*Overall project to develop the software*	
User Interface	Design and develop the user interface	Design login screen
		Create navigation menu
		Implement user feedback
		Test user interface
Database Design	Plan and create the database structure	Define database schema
		Establish data relationships
		Develop data validation rules
		Perform database testing
Application Logic	Build the core application logic	Develop business logic
		Implement algorithms
		Integrate third-party libraries
		Conduct unit testing

Table 5 - Software Application WBS – Outline Style

The table above provides a hierarchical breakdown of the major tasks for developing a software application, including higher-level tasks like User Interface, Database Design, and Application Logic, which are broken down into more specific subtasks. This structured approach helps in planning, organizing, and tracking the project's progress.

Work Package Breakdown Structure (WPBS)

The Work Package Breakdown Structure (WPBS) or (PBS), also known as the Package Register, is a crucial component of project management. For projects that have failed, this is one the WBS that is normally left out and is not controlled and managed on an ongoing basis. A project that does not have a Work Package Breakdown Structure incorporated usually slips on their time and obviously costs as well. This PBS provides a hierarchical breakdown of all the work packages or deliverables that must be completed for a project. The PBS helps organize and structure the project by breaking it down into manageable and smaller components.

The PBS serves as a framework for defining the scope of work and ensures that all the necessary tasks are identified and assigned to the appropriate team members. It helps in understanding the dependencies between different packages and identifies the relationships between them. This allows for better coordination and integration of the project activities.

The PBS also serves as a communication tool to the project team regarding the procurement items of a project, as it provides a clear and concise overview of the project's structure and deliverables. It helps stakeholders and team members understand the overall project and their specific roles and responsibilities. This transparency enhances collaboration and ensures that everyone is aligned with the project objectives.

Furthermore, the PBS enables effective project planning and scheduling. By breaking down the project into smaller packages, estimating the resources, time, and cost required for each package becomes easier. This allows for more accurate project planning and helps track and monitor progress against the defined packages.

Overall, the Work Package Breakdown Structure is a crucial tool in project management. It provides a structured approach to organizing and managing projects, ensuring that all deliverables are identified, assigned, and completed. It enhances communication, coordination, and collaboration among team members and stakeholders, ultimately leading to successful project outcomes.

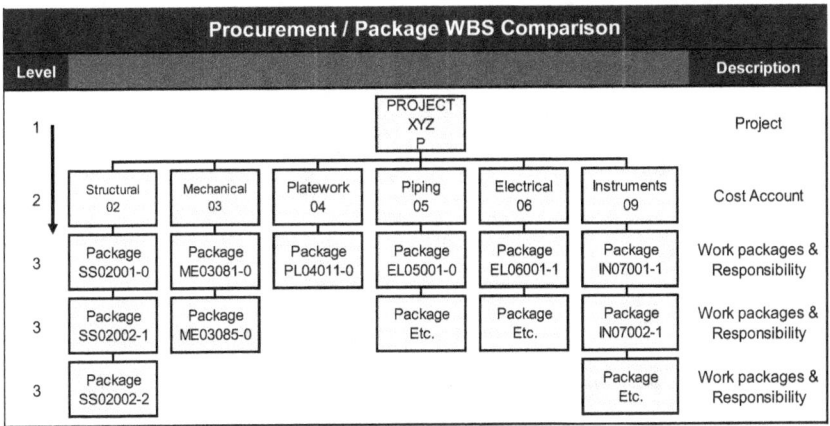

Figure 6 - Sample Package / Procurement WBS

Systems Breakdown Structure (SBS) (for Commissioning)

The systems breakdown structure (SBS) is a critical tool used during a project's commissioning and handover phase. It provides a comprehensive breakdown of all the systems and components within a facility, allowing for easier management and oversight during the commissioning process.

The project team typically creates the SBS and includes detailed information about each system, such as its purpose, function, and interdependencies with other systems. This structure is essential for ensuring that all systems are properly integrated and functioning as intended before the facility is handed over to the end user.

During the commissioning phase, the SBS serves as a roadmap for testing and verifying the performance of each system. Clearly defining each

45

system's scope and boundaries helps identify any potential issues or conflicts that may arise during the commissioning process. This allows the project team to address these issues proactively, minimizing delays and reducing the risk of costly rework.

Additionally, the SBS is a valuable reference tool for the facility owner or operator during the handover phase. It provides a clear overview of the facility's systems, making it easier to understand and manage the operation and maintenance of the facility. This information is particularly important for training staff and ensuring they comprehensively understand the facility's systems and interdependencies.

Overall, the systems breakdown structure is crucial to the commissioning and handover process. It helps to ensure that all systems are properly integrated and functioning as intended, minimizing the risk of errors or inefficiencies. By providing a clear overview of the facility's systems it also facilitates effective management and operation of the facility once it is handed over to the end user.

NOTE: The biggest mistake project teams make is to start the planning for the commissioning and handover phase much too late during the planning or execution phase. If you only start thinking of setting up commissioning systems during execution, you have not planned for commissioning and handover, and you will run into progress and delay issues. I can guarantee it!

Systems are subsets of the project that you can start testing before startup and ensure that they are running smoothly. Most projects can break their WBS into system packs as well. Start delivering in small your project in small chunks.

*It will assist projects with a product (e.g., a Mine
(deliver ore), Oxygen plant, Sugar Mill, etc.) that must
be tested or delivered. System issues typically happen
during "hot commissioning" or "C3".*

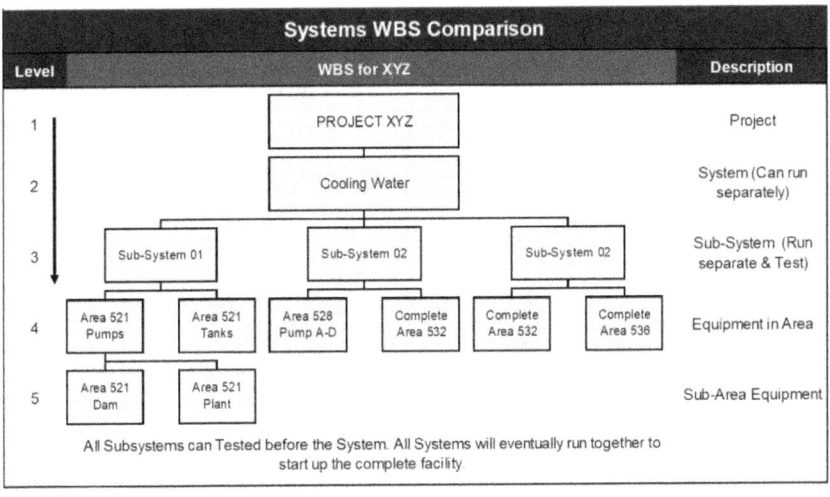

Figure 7 - Simple Systems WBS used for Start=up

Estimating Breakdown Structure (EBS)

The Estimating Breakdown Structure (EBS) is a project management tool used to organize and categorize a project's various components to estimate costs and resources. Similar to a Work Breakdown Structure (WBS), the EBS breaks down a project into smaller, more manageable pieces. However, instead of focusing on the work packages or tasks, the EBS focuses on the cost and resource elements of the project.

The EBS is typically created during the project planning phase and is the foundation for estimating project costs and resources. It provides a hierarchical structure that allows project managers to identify and categorize

47

the various cost and resource elements required for project execution.

The EBS is composed of multiple levels, with each level representing a different level of detail. At its highest level, the EBS may include major categories, such as labor, materials, equipment, and overheads. As the structure progresses to lower levels, these categories are further broken down into more specific subcategories. For example, under the labor category, there may be subcategories for different types of labor, such as skilled labor, unskilled labor, or specialized labor.

The EBS provides a more accurate representation of the project's requirements by breaking down the project into smaller cost and resource elements. This allows for more accurate cost and resource estimates and better resource allocation and budget planning.

The EBS also serves as a communication tool, allowing project stakeholders to easily understand and review each project element's estimated costs and resources. It provides a clear, organized structure that can be easily communicated and shared with team members, clients, and other stakeholders.

In addition to cost and resource estimation, the EBS can be used to track and control project costs. By comparing the actual costs incurred during project execution to the estimated costs in the EBS, project managers can identify any cost variances and take appropriate actions to address them.

Overall, the Estimating Breakdown Structure is a valuable project management tool that helps in accurately estimating project costs and resources. It provides a structured and organized approach to cost estimation, allowing project managers to effectively plan, track effectively, and control project costs throughout the project lifecycle.

NOTE: On bigger projects, I always insist that the

estimating team develop an Estimating Breakdown Structure (EBS) to assist the project team in aligning with a few items: One. the Work Package Breakdown Structure (WPBS). Two, the Cost Breakdown Structure (CBS). The third is the Schedule Breakdown Structure (SBS). When all these are aligned, your project control should be a breeze. See the example below.

Here is a typical list that quantity surveyors use to produce an estimate. These types of reports are used in all project spheres and industries. In this specific case, I will use a house to demonstrate the various interfaces and how that will affect your ability to keep track of change control and manage your packages and costs simultaneously.

Remember, various structures interface with each other in multiple ways. The ratios between the various WBS could be as follows, not limited to:

One (1) to Many: e.g., WBS Element > Schedule Activities.

Many to Many: e.g., Equipment (Pumps) = Schedule Activities.

Many to One (1): e.g., Estimate Items -> Procurement Package.

One (1) to One (1): e.g., Estimate Account -> Cost Account

The following table shows this project's Masonry bill of quantities (BOQ). This QS structure is not good enough because when changes are made, it is very difficult to change the quantities as you cannot just remove a section or portion of the work. This estimate is done as per WBS structure in table 4. We will demonstrate a WBS structure where change control is simplified by allocating WBS or EBS structures to the BOQ.

Dunes Guest House Sample BOQ					VG
Item	Description	Unit	Quantity	Rate	Amount
	MASONRY				
	FOUNDATIONS (PROVISIONAL)				
	Brickwork of 14 Mpa cement bricks in class I mortar				
1	Half brick walls	m²	11.04	179.18	1,978.15
2	One brick walls	m²	68.31	358.50	24,489.14
3	One and half brick walls	m2	6.28	537.68	3,376.63
	SUPERSTRUCTURE				
	Brickwork of 7 Mpa cement bricks in class II mortar				
4	Half brick walls	m²	121.57	166.43	20,232.90
5	One brick walls	m²	271.52	333.01	90,418.88
	Brickwork of 14 Mpa cement bricks in class I mortar				
6	Half brick walls	m²	46.36	179.18	8,306.78
7	One brick walls	m²	293.83	356.07	104,624.05
8	One and half brick walls	m2	16.37	537.68	8,801.82
	BRICKWORK SUNDRIES				
	Brickwork reinforcement				
9	75mm Wide reinforcement built in horizontally	m	558.43	1.26	703.62
10	75mm Wide reinforcement built in horizontally in foundations (Provisional)	m	157.25	1.26	198.14
11	150mm Wide reinforcement built in horizontally in foundations (Provisional)	m	678.03	1.29	874.66
12	150mm Wide reinforcement built in horizontally	m	1,787.40	1.29	2,305.75
	Prestressed fabricated lintols				
13	110 x 70mm lintols in lengths not exceeding 3m	m	170.59	73.30	12,504.25
	FACE BRICKWORK				
	Rockface bricks (40MPa) in class II mortar pointed with recessed horizontal and vertical joints				
14	Extra over brickwork for face brickwork in soldier courses in foundations (Provisional)	m²	36.01	104.79	3,773.49
	STONE WALLING				
	Natural stone walling of uncoursed random stone, including facing and pointing on one side including chipping sizing of stones and natural stone sealer to surface when completed				
15	110mm Linings to brickwork, including wire ties (Brickwork measured elsewhere)	m²	121.71	867.88	105,629.67
	MASONRY				**388,217.93**

Table 6 - Estimating Typical QS WBS Coding

Here is a sample guest house WBS / EBS. Your overall or other WBSs for this project might be tracked at a higher or lower level. For example, The *Tandem Garage* and the *Jacuzzi & Gym* areas are planned to be built right above each other in this guest house. By coding the structure of the estimate as per the example below, we can remove both the *Tandem Garage* and the *Jacuzzi & Gym* areas by deleting the two WBS / EBSs.

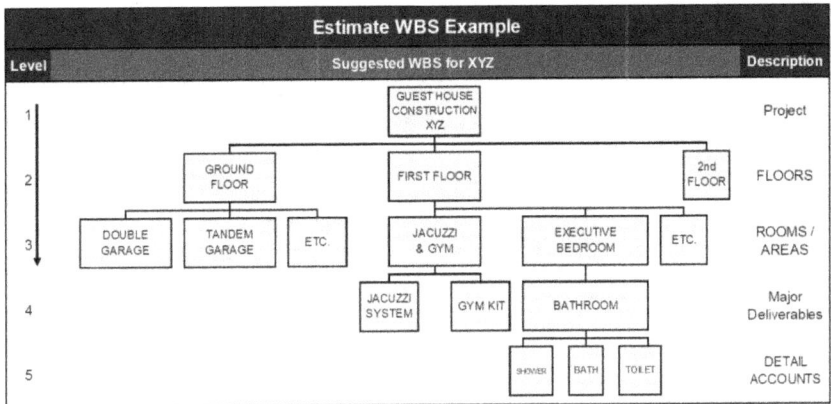

Figure 8 - Estimating Sample - Guest House EBS

The figure below depicts the highlighted areas that are removed.

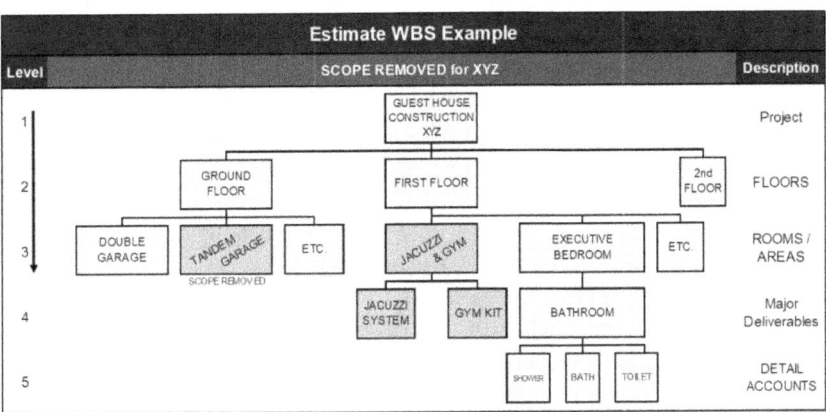

Figure 9 - Estimating - Sample Guest House EBS (Scope Removed)

Removing these two areas in level 3 of the EBS will change the estimate and the EBS in one go.

51

Level 1	Level 2	Level 3	Level 4
Construction of Guest House - Entire Project			
	Ground Floor	Double Garage	
		Tandem Garage	
		Stairs	
		Guest Toilet	Toilet
			Handbasin
		Entertainment Area	Bar
			Fireplace
		Lounge Area	
		Kitchen Area	Pantry
			Scullery
			Laundry Room
	First Floor	Jacuzzi & Gym area	Jacuzzi
			Gym kit
		Executive Bedroom	Bathroom
			Toilet
		Etc.	

Table 7 - Estimating - Sample GUEST HOUSE Outline

We have not included the complete WBS of the Guest house here, but it will be quite detailed if you present the full estimate structure. This will align with the section in Part III, where we explain the steps to follow to produce a detailed bottom-up estimate as part of this book.

Schedule Breakdown Structure (SBS)

The Schedule Breakdown Structure (SBS) is a hierarchical representation of the tasks and activities involved in a project schedule. Similar to the Work Breakdown Structure (WBS), which breaks down project deliverables into smaller, manageable components, the SBS focuses specifically on breaking down the project schedule into smaller, more manageable time frames.

The SBS serves as a visual tool that helps project managers and team members understand the sequence and dependencies of activities within the project schedule. It provides a clear and organized structure for scheduling and tracking project activities, allowing for effective time management and resource allocation.

At the highest level, the SBS typically represents the major phases or stages of the project. Phases can be divided into smaller time intervals, such as months, weeks or days, depending on the project's duration and complexity. Each time interval is then broken down into specific activities or tasks that must be completed within that timeframe.

The SBS can be displayed in various formats, including a Gantt chart or a timeline, to visually represent the schedule breakdown. Within each time interval or activity, dependencies and relationships are identified to ensure that tasks are sequenced correctly and that any dependencies are accounted for. Usually, the client's team WBS will be at a higher level than the construction team and is set up so that it can roll up the information to the next level of the WBS. See the difference between the deliverable-based WBS and the schedule breakdown structure.

One of the main benefits of using the SBS is that it allows project managers to easily identify critical paths and potential bottlenecks in the schedule. Breaking down the project schedule into smaller time intervals makes it easier to identify tasks critical to the overall project timeline. This enables project managers to prioritize and allocate resources accordingly, ensuring that the project stays on track and meets its deadlines.

Moreover, the SBS also helps in resource allocation and workload distribution. Breaking down the schedule into smaller intervals and tasks makes it easier to estimate the effort required for each task and allocate resources accordingly. This ensures that team members are not overloaded with work and that resources are utilized efficiently.

The Schedule Breakdown Structure (SBS) is a valuable tool for project managers to effectively manage and track project schedules. Breaking down the schedule into smaller intervals and tasks allows for better time management, resource allocation, and identification of critical paths. It visually represents the project schedule, making it easier for project teams to plan and execute their work effectively.

NOTE: A Schedule Breakdown Structure (SBS) should be developed by your implementation project team that will execute the construction portion of a project. Normally, an Owner's team or Client's Team (OT) has a WBS structure that goes maybe 4 to 5 levels deep to control their project. However, to assist the OT with control, the construction team requires a lower level of control to manage the deliverables and keep the progress on track.

When you encounter a troubled project and want to turn it around, it will assist your project team in keeping a finger on the pulse of your project.

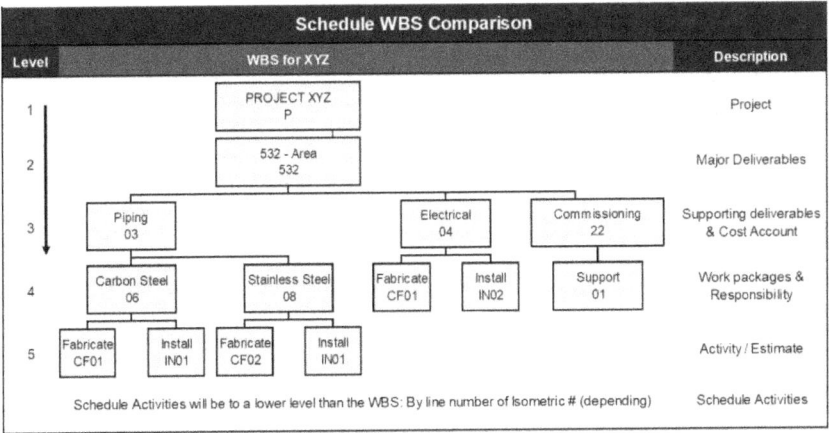

Figure 10 - Sample Schedule WBS

Level	WBS Number	#	Description	Detail
1	X	X	Project XYZ	Constructioin portion of Project XYZ
2	X.532	532	Area 532	
3	X.532.03	03	Piping	Piping Fabrication & Installlation
4	X.532.03.06	06	Carbon Steel	Carbon Steel Fabrication & Installlation
5	X.532.03.06.CF01	CF01	CS Fabrication	Carbon Steel Piping Fabrication
5	X.532.03.06.IN01	IN01	CS Insatllation	Carbon Steel Piping Installlation
4	X.532.03.06	08	Stianless Steel	Stainless Steel Fabrication & Installlation
5	X.532.03.06.CF01	CF02	SS Fabrication	Stainless Steel Piping Fabrication
5	X.532.03.06.IN01	IN01	SS Insatllation	Stainless Steel Piping Installlation
	etc.			

Table 8 - Schedule WBS Coding

Cost Breakdown Structure (CBS)

The Cost Breakdown Structure (CBS) is a hierarchical representation of the costs associated with a project or a business. It provides a detailed breakdown of all costs, allowing for better cost estimation, budgeting, and financial control.

The CBS is created by breaking down the project or business into smaller components, activities, or work packages. Each component is then further

55

divided into its cost elements, allowing for a granular view of the costs. The structure typically starts at a high-level and progressively drills down into more detailed levels.

The CBS serves several purposes. Firstly, it helps estimate costs accurately by identifying all cost elements and clearly understanding what needs to be budgeted for. Breaking down costs into smaller components eliminates the chances of overlooking any expenses.

Secondly, the CBS helps project teams in allocating budgets and resources. It allows project managers or business owners to allocate funds to specific cost elements or work packages based on their priority and importance. This helps in effective financial planning and ensures that adequate resources are available for each component.

Furthermore, the CBS provides a basis for tracking and controlling costs throughout the project or business lifecycle. By having a detailed breakdown of costs, any deviations or variances can be easily identified and rectified. It enables effective cost management by comparing actual costs against the budgeted amounts and taking corrective actions if required.

The CBS also facilitates better decision-making. With a clear understanding of the cost breakdown, project managers or business owners can make informed decisions regarding resource allocation, cost-saving measures, and investment priorities. It helps identify areas where cost optimization can be achieved and also facilitates effective cost control.

Additionally, a CBS serves as a communication tool. It allows stakeholders, project teams, or management to understand the cost and breakdown structure. It ensures transparency and clarity in financial discussions and reporting, enabling effective communication between parties.

In conclusion, the Cost Breakdown Structure (CBS) is a valuable tool for cost estimation, budgeting, financial control, and decision-making. It provides a detailed breakdown of costs, allowing for accurate estimation, effective resource allocation, and improved cost management. By facilitating transparency and communication it enables stakeholders to have a clear understanding of the cost structure and supports informed decision-making.

NOTE: This Cost Breakdown Structure (CBS) is required for all projects, and it also ties in with the following: One, the Schedule (normally the WBS) and two, the Estimate. It is possible to have a Schedule Breakdown Structure (SBS) different from the WBS. The Schedule (SBS) and the Estimate (EBS) normally have more detail than the WBS.

The CBS can (should) align with the different structures and deliverables of the project. If set up correctly, then Scope & Change Control is easy.

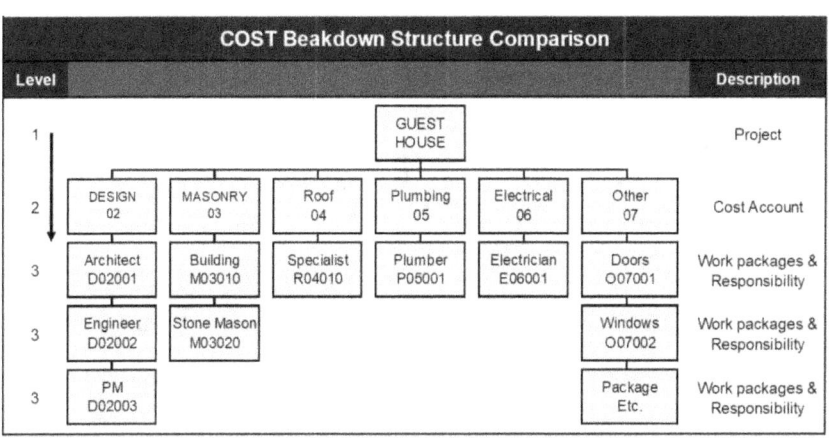

Figure 11 - Costing - Sample Guest House CBS

Key Tips: Cardinal Breakdown Structures

These cardinal breakdown structures should be used to ensure the success of any project. I normally have all of these as part of my overall structures that help me manage various project phases.

Here are some key tips I have learnt about the need for the cardinal or important breakdown structures to consider:

Plant Breakdown Structure (PBS)

First and foremost, with my extensive experience with plant construction and management, I've come to rely on the Plant Breakdown Structure (PBS) as a fundamental framework. It serves as the backbone for dissecting the complex anatomy of a plant into a clear, manageable structure.

- **Cataloging of Components:** The PBS provides a detailed inventory of all components, systems, and subsystems within a plant, aligned with the System Breakdown Structure for detailed analysis.

- **Hierarchical Segmentation:** At the highest level, the plant is divided into primary sections including the main building, processing units, and utilities.

- **Detailed Dissection:** Each primary section is further broken down into finer elements such as machinery and control systems.

- **Comprehensive Mapping:** This detailed mapping is crucial for understanding the plant's complexities and interdependencies, ensuring a comprehensive approach from design to maintenance.

- **Clarification of Interrelations:** The structured layout of the PBS clarifies the relationships between components, streamlining management processes.

- **Facilitation of Communication:** It enhances communication across teams, ensuring everyone is on the same page.

- **Project Management Guidance:** The PBS is vital for guiding key project management tasks, including cost forecasting, scheduling, risk evaluation, and resource allocation.

- **Foundation for Work Scope Breakdown:** It ensures a thorough breakdown of work scope, identifying all critical activities and dependencies.

In conclusion, the PBS transcends its role as a mere tool; it is a strategic partner throughout the lifecycle of a plant. It brings clarity and precision to the complex process of plant management, ensuring that every phase of the project, from inception to completion, is marked by efficiency and thoroughness. The PBS is integral to my methodology, ensuring that project execution is not only streamlined but also strategically aligned with the project's long-term success.

Work Package Breakdown Structure (PBS)

The Work Package Breakdown Structure (WPBS), also known as the Package Register, is an essential tool in my project management arsenal, serving as the blueprint for organizing project deliverables into a clear and manageable hierarchy.

- **Structured Hierarchy:** Establish a structured hierarchy of project deliverables using the PBS.

- **Segmentation into Packages:** Break down the project into smaller, manageable packages for precise scope definition and delegation.

- **Definition and Delegation:** Meticulously define each project facet and delegate to the right team members.

- **Illumination of Interdependencies:** Highlight the interdependencies among packages to enhance coordination and task integration.

- **Communication Tool:** Utilize the PBS as a communication platform to provide stakeholders and team members with a clear view of the project structure and roles.

- **Project Planning and Scheduling:** Support precise project planning and scheduling by enabling accurate estimations of resources, time, and budget for each package.
- **Tracking and Monitoring:** Facilitate vigilant tracking and progress monitoring through detailed package breakdown.

The PBS is more than just a framework; it's a strategic approach that ensures every aspect of the project is accounted for and executed with precision. By breaking down projects into defined packages, it not only enhances the clarity of the project scope but also improves stakeholder communication and team collaboration. The PBS's role in facilitating detailed planning and resource allocation cannot be overstated, as it allows for close monitoring of each project phase. Ultimately, the PBS is indispensable for successfully delivering projects, ensuring that every deliverable is managed efficiently.

Systems Breakdown Structure (SBS)

I have extensive experience in overseeing project commissioning, handover and ramp-up, and the *Systems Breakdown Structure (SBS)* has been pivotal during those stages. This tool meticulously catalogs all systems and components within a facility, streamlining management and ensuring thorough oversight throughout commissioning. Crafted by the project team, the SBS details each system's purpose and interrelations, which is crucial for seamless integration and optimal functionality at handover.

As a commissioning blueprint, the SBS guides the validation of each system's performance, clarifying the scope and pinpointing potential discrepancies early on. This foresight is critical in preempting delays and mitigating the need for expensive corrective measures. Moreover, the SBS is invaluable during handover, offering the facility's future custodians a lucid understanding of the operational landscape, which is essential for effective maintenance and staff training.

A cardinal rule I advocate is the early incorporation of commissioning planning within the project lifecycle. Delaying this until the execution phase is a recipe for setbacks and delays. Systems are, after all, project subsets that warrant early testing to ensure smooth startup. Integrating system packs into the Work Breakdown Structure (WBS) allows for the project to be delivered in manageable increments, ensuring each segment is up to par before moving on to the next. In essence, the SBS is not just a tool but a strategic framework, ensuring that each system is robustly vetted and ready for operation, thereby guaranteeing a smooth transition to the end user.

Estimate breakdown structure (EBS)

The Estimate breakdown structure (EBS) is essential in providing a realistic and accurate assessment of the project's time and cost requirements. The EBS is a methodical approach to dissecting a project into smaller parts, which is essential for precise estimations of time, effort, and resources. This process has imparted several key lessons that are integral to successful project management.

- **Thorough Planning and Analysis:** Emphasize careful analysis of each project component, considering risks and dependencies.
- **Stakeholder Collaboration:** Involve team members and experts for diverse insights and enhanced estimate accuracy.
- **Strive for Accuracy:** Utilize historical data and benchmarks to refine estimates for realistic planning and expectation management.
- **Flexibility and Adaptability:** Maintain readiness to adjust estimates in response to project changes or new information.
- **Iterative Process:** Treat estimating as an ongoing activity, with regular reviews and updates as the project evolves.

The practice of breaking down estimates teaches us the value of meticulous planning and the collective wisdom of stakeholder collaboration. It underscores the need for accuracy, yet reminds us to remain flexible and adaptable as project landscapes change. Recognizing that this is not a one-off task but an iterative process ensures that our projects stay on course and adapt to new challenges. By internalizing these lessons, we can enhance our project outcomes and optimize our use of resources, leading to a more dynamic and responsive approach to project management.

Scheduling breakdown structures (SBS)

The Scheduling breakdown structures (SBS) are a cornerstone in project management, offering a systematic approach to task organization and planning. My journey in project management has imparted several key insights into the effective use of these structures.

- **Task Decomposition:** Break the project into smaller, manageable tasks for clearer allocation of resources, time estimation, and progress tracking.
- **Dependency Clarity:** Define clear dependencies between tasks to understand execution order and identify potential bottlenecks.
- **Dynamic Review:** Continuously review and update the structure to accommodate project evolution, shifting priorities, and emerging challenges.
- **Team Involvement:** Involve the project team in developing the structure to ensure accuracy and foster a sense of ownership and accountability.
- **Resource Allocation:** Consider the availability and expertise of resources to ensure tasks are assigned efficiently and effectively.
- **Progress Monitoring:** Regularly track task progress against the structure to identify deviations and implement timely corrective actions.

In synthesizing these experiences, it's evident that scheduling breakdown structures are not static but dynamic tools that require active management and adaptation. They are the scaffolding upon which a project's success is built, providing a blueprint for meticulous planning and execution. By breaking down projects into smaller components, clearly defining task dependencies, and ensuring regular updates and team involvement, these structures become a guide for resource allocation and progress monitoring. The lessons learned from applying scheduling breakdown structures reinforce their significance in steering projects toward their timely and successful completion.

NOTE: It is a pity (or in my mind, sacrilege) that most review teams refer to the SBS as the WBS of the project. This should be a big red flag on all your projects!

To the Review Teams and the Reviewing Consultants: Shame on you!

Cost Breakdown Structure (CBS)

The Cost Breakdown Structure (CBS) plays in the meticulous management of project finances. The CBS is a tool that dissects project costs into a detailed hierarchy, which is essential for precise cost estimation, informed budgeting, and diligent tracking of expenses as a project unfolds.

- **Hierarchical Cost Breakdown:** Implement a CBS to provide a structured overview of all project costs.
- **Accurate Estimations:** Use the CBS for precise cost estimations to prevent budget overruns.
- **Budgeting and Tracking:** Leverage the CBS for thorough budgeting and continuous tracking of expenses.
- **Categorization of Costs:** Categorize costs within the CBS by components such as hardware, software licenses, labor, and

training.

- **Cost-Saving Identification:** Analyze the CBS to identify and implement cost-saving measures.

- **Optimization of Resources:** Use insights from the CBS to optimize resource utilization, like software licenses, for cost efficiency.

- **Integration with Other Structures:** Combine the CBS with other breakdown structures like the OBS (Organizational Breakdown Structure) and RBS (Risk Breakdown Structure) for comprehensive project management.

The implementation of a CBS in a challenging software development project underscored its value. It was a turning point that allowed us to gain control over spiraling costs and implement strategic cost-saving measures. This experience cemented my belief in the necessity of integrating various breakdown structures, such as the CBS, alongside the WBS, OBS, and RBS to form a robust framework for project management. These structures are tracking tools and are instrumental in providing insights, managing risks, engaging stakeholders, and ensuring financial reliability. By embracing these structures, project managers are equipped to steer projects toward successful outcomes with financial precision and strategic foresight.

Conclusion

In conclusion, using these cardinal breakdown structures in addition to the WBS is crucial for successful project management, especially during execution and construction. These structures provide a systematic approach to managing various aspects of a project, ensuring better control, communication, and, ultimately, the successful delivery of the project. As a project manager, I highly recommend incorporating these breakdown structures into your project management toolkit for optimal results. By incorporating these breakdown structures into my project during planning and execution, I am able to enhance project outcomes and ensure its overall success.

NOTE: Not all the breakdown structures have to be used, and the project team must utilize these structures to manage their own areas for better control.

Lessons Learnt: Cardinal Breakdown Structures

Plant or Facility Construction

Plant Breakdown Structures (PBS) have proven to be invaluable tools for planning, organizing, and managing projects in various industries. Through my experience, I have learned several lessons that highlight the importance and benefits of utilizing PBS.

Firstly, one crucial lesson learnt is that a **PBS provides a systematic and hierarchical framework** for breaking down complex projects into manageable components. Composing the project into smaller work packages makes it easier to assign responsibilities, estimate resource requirements, and develop planning schedules for a project. This level of organization helps prevent project delays and ensures that all tasks are accounted for and executed efficiently.

Another valuable lesson is that the PBS **promotes effective communication and collaboration** among project stakeholders. The clear structure allows for better understanding and alignment of project objectives, deliverables, and dependencies. This transparency helps to minimize misunderstandings, conflicts, and rework. It also facilitates effective decision-making and enables stakeholders to track progress and evaluate project performance accurately.

Furthermore, the PBS **aids in identifying and managing project risks**. By breaking down the project into its constituent parts, potential risks can be identified at each level, and appropriate mitigation strategies can be implemented. This proactive approach to risk management ensures that potential issues are addressed in a timely manner, reducing the likelihood of costly delays or failures.

In addition, the PBS **fosters a culture of accountability and ownership** among project team members. Each work package within the PBS is assigned to a specific individual or team, making it clear who is responsible for its completion. This clarity of roles and responsibilities promotes a sense of ownership and encourages individuals to take pride in their work, leading to increased productivity and quality output.

Lastly, the PBS serves as a **valuable reference tool for future projects**. By documenting the breakdown structure and associated information, organizations can build a repository of knowledge and lessons learned. This repository can be used to improve project planning, streamline processes, and enhance overall project management capabilities.

In conclusion, lessons learnt from working with Plant Breakdown Structures emphasize their significance in project management. PBS provides a structured approach to project planning, enhances communication and collaboration, helps manage risks, fosters accountability, and is a valuable resource for future projects. Through implementing a PBS, organizations can improve project outcomes, reduce costs, and achieve greater overall success.

Developing a Product

One of the key lessons learned about Product Breakdown Structures (PBS) is **the importance of proper planning and organization**. Developing a PBS requires a deep understanding of the product and its components. It is crucial to invest time and effort in breaking down the product into smaller, manageable components, ensuring that nothing is overlooked. This planning phase helps identify all the necessary tasks, resources, and dependencies associated with each component.

Another lesson learned is **the significance of effective communication**. Developing a PBS often involves collaboration among various stakeholders, including product managers, engineers, designers, and other team members. Clear and concise communication is essential to ensure everyone understands the breakdown structure and their respective roles and responsibilities. Regular meetings and discussions help to refine and update the PBS as needed, ensuring everyone is aligned and on the same page.

A key lesson learned is **the value of flexibility and adaptability**. As projects progress, there may be changes in requirements, scope, or priorities. A well-designed PBS allows for easy adjustments and modifications without disrupting the overall project structure. It is important to regularly review and update the PBS to incorporate any changes and keep it relevant and accurate.

Additionally, it is essential to **involve subject matter experts throughout the development of a PBS**. Their expertise and knowledge contribute to a more comprehensive product breakdown and ensure that all relevant components are included. Their input also helps identify potential risks, challenges, and dependencies, enabling better planning and mitigation strategies.

Lastly, a lesson learned is **the need for continuous monitoring and control**. Once the PBS is established, it serves as a framework for tracking progress, managing resources, and evaluating the overall project performance. Regular monitoring helps identify any deviations from the original plan and promptly take corrective actions. It allows for better control over the project's timeline, budget, and quality, ultimately leading to successful product delivery.

In conclusion, developing a Product Breakdown Structure requires careful planning, effective communication, flexibility, expert involvement, and continuous monitoring. Incorporating these lessons learned can significantly enhance the efficiency and success of project management and product development.

Managing Procurement Packages

One important lesson learned from Package Breakdown Structures is the significance of **proper planning and organization**. Creating a detailed breakdown structure helps identify all the necessary components and tasks required for a project. This enables better planning and allocation of resources, ensuring a smooth workflow and minimizing the risk of overlooked tasks or requirements.

Another lesson learned is the **importance of clear communication and collaboration**. Package Breakdown Structures involve multiple stakeholders responsible for different packages or components. Establishing effective communication channels and fostering collaboration among team members is crucial to ensure that everyone understands their roles and responsibilities. This promotes transparency, reduces misunderstandings, and facilitates successful project completion.

Package Breakdown Structures also teach the lesson of **adaptability and flexibility**. As projects progress, there may be changes in requirements, priorities, or resource availability. Having a well-defined breakdown structure allows for easier identification and adjustment of packages or tasks affected by these changes. The ability to adapt and make necessary adjustments ensures that the project remains on track and can accommodate any modifications without major disruptions.

Another lesson learned from Package Breakdown Structures is the **importance of monitoring and tracking progress**. A comprehensive breakdown structure enables project managers to track the completion of each package or task, ensuring that deadlines are met and milestones are achieved. Regular monitoring and progress tracking help identify any potential bottlenecks or delays, allowing for timely interventions and adjustments to keep the project on schedule.

Lastly, Package Breakdown Structures emphasize the lesson of **accountability and responsibility**. Each package or component is assigned to a specific individual or team, making them accountable for its successful completion. This promotes a sense of ownership and responsibility, encouraging individuals to take pride in their work and deliver quality

69

results. It also enables efficient problem-solving, as responsible parties can be identified quickly in case of any issues or challenges.

In conclusion, Package Breakdown Structures provide valuable lessons in terms of planning, communication, adaptability, monitoring, and accountability. By effectively applying these lessons, project managers can enhance project success rates and ensure efficient and effective project execution.

Commissioning & Handover Guidance

One important lesson learned from systems breakdown structures is the necessity of thorough **planning and organization**. Creating a comprehensive breakdown structure allows for a clear understanding of the different components of a system and how they are interconnected. Identifying and addressing potential breakdown system points becomes challenging without proper planning and organization.

Another lesson learned is the significance of **communication and collaboration among team members**. Systems breakdown structures require input from various stakeholders, such as plant managers, engineers, designers, and project managers. Effective communication ensures that everyone is on the same page regarding the breakdown structure, allowing for better coordination and problem-solving when breakdowns occur. Collaboration also helps identify areas of improvement and find innovative solutions to prevent breakdowns in the future.

Systems breakdown structures also highlight the **importance of risk analysis and mitigation**. Breaking down a system into its components makes identifying potential risks and vulnerabilities easier, especially in the sub-systems. This analysis enables proactive measures to be taken to mitigate these risks and prevent breakdowns. It also helps develop contingency plans and backup systems to ensure minimal disruptions during a breakdown.

One crucial lesson learned is the need for **continuous monitoring and maintenance of systems**. Breakdowns can occur due to various reasons, such as wear and tear, outdated technology, or external factors. Regular monitoring and maintenance allow for the early detection of potential breakdown points and prompt actions to rectify them. This lesson emphasizes the importance of investing in regular inspections, upgrades, and maintenance to ensure the optimal functioning of the system and prevent major breakdowns.

Lastly, systems breakdown structures teach us the **importance of adaptability and resilience**. Despite careful planning and risk mitigation strategies, breakdowns can still occur unexpectedly. Being adaptable and resilient in such situations is crucial to minimize the impact of breakdowns and quickly restore normal operations. This lesson emphasizes the need for contingency plans, backup systems, and the ability to quickly respond and recover from breakdowns, ensuring the system's overall stability and reliability.

Estimating Breakdown Structures

Estimating breakdown structures refers to breaking down a project or task into smaller, more manageable components to estimate the time, effort, and resources required for its completion. Through this process, several valuable lessons can be learned.

One lesson is the importance of **thorough planning and analysis**. Estimating breakdown structures requires a deep understanding of the project and its requirements. It is crucial to carefully analyze each component and consider any potential risks, dependencies, or constraints that may impact the estimation. Taking the time to plan and analyze upfront can help avoid inaccuracies and uncertainties later on.

Another lesson is the need for **collaboration and input from various stakeholders**. Estimating breakdown structures should not be done in isolation. Involving team members, subject matter experts, and other relevant stakeholders can provide valuable insights and perspectives that may enhance the accuracy and reliability of the estimates. Collaboration also fosters a sense of ownership and accountability among the team, leading to better overall project outcomes.

Accuracy is another key lesson learned in estimating breakdown structures. It is essential to strive for as much accuracy as possible in the estimates. This involves considering historical data, benchmarking against similar projects, and continuously refining the estimation process. Ensuring accuracy in estimates helps in realistic project planning, resource allocation, and managing stakeholders' expectations.

Additionally, **flexibility and adaptability** are lessons learnt in estimating breakdown structures. Projects and tasks may evolve or change over time, and estimations may need to be adjusted accordingly. Being open to revisiting and recalibrating the breakdown structures allows for more accurate estimations, even when faced with unforeseen circumstances or changes in project scope.

Lastly, it is important to recognize that estimating breakdown structures is not a one-time activity. It is an iterative process that requires continuous monitoring, evaluation, and improvement. Regularly **reviewing and updating estimations** as the project progresses helps identify deviations or variances from the initial estimates and allows for timely corrective actions.

In conclusion, estimating breakdown structures is a valuable practice that provides insights into project planning, resource allocation, and risk management. Lessons learnt from this process include the significance of thorough planning and analysis, collaboration with stakeholders, accuracy in estimates, flexibility and adaptability, and the need for continuous improvement. Applying these lessons can lead to more successful project outcomes and effective resource utilization.

Scheduling Breakdown Structures

Scheduling breakdown structures play a crucial role in project management, helping to **organize and plan tasks effectively**. Through my experience, I have learned several valuable lessons about these structures.

Firstly, breaking down the project into **smaller, manageable tasks** is essential. Dividing the project into smaller components makes allocating resources, estimating timeframes, and tracking progress easier. This breakdown provides clarity and ensures that all aspects of the project are accounted for.

Another lesson learned is the **importance of defining clear dependencies between tasks**. Understanding the interdependencies helps determine the order in which tasks must be executed and identifies any potential bottlenecks or constraints. Project managers can minimize delays and optimize resource allocation by establishing these dependencies.

Regularly **reviewing and updating the scheduling breakdown structure** is another lesson gained from experience. As projects evolve, new tasks may arise, priorities may shift, and unforeseen challenges may emerge. Therefore, it is crucial to continuously assess and adjust the scheduling breakdown structure to reflect the current status and ensure that the project remains on track.

Additionally, **involving the entire project team in developing and understanding** the scheduling breakdown structure is essential. Collaboration and input from team members contribute to a more accurate and comprehensive structure. This involvement also fosters a sense of ownership and accountability among team members, leading to increased commitment and efficiency.

Furthermore, it is crucial to consider the **availability and expertise of resources** when creating the scheduling breakdown structure. Assigning tasks to individuals with the appropriate skills and availability ensures that work is completed efficiently and proper quality. Overloading or underutilizing resources can lead to delays or inefficiencies, impacting the overall project timeline.

Lastly, **monitoring and tracking the progress of tasks** against the scheduling breakdown structure is vital. Regularly reviewing the actual progress against the planned schedule helps identify any deviations or potential issues early on. This allows for timely adjustments and corrective actions, improving the chances of meeting project deadlines.

In conclusion, scheduling breakdown structures are integral to successful project management. Breaking down projects into smaller tasks, defining dependencies, regularly reviewing and updating the structure, involving the entire team, considering resource availability, and monitoring progress are all valuable lessons learned from experience. By applying these lessons, project managers can enhance their ability to plan, execute, and deliver projects efficiently and effectively.

Effective Cost Management

The significance of utilizing the Cost Breakdown Structure (CBS) to manage project costs effectively will dawn on you if you do not manage costs on a CBS. Most companies have an accounting system that tracks the actual costs. The CBS provides a hierarchical breakdown of project costs, allowing for accurate estimation, budgeting, and tracking of expenses throughout the project lifecycle.

In a recent training development project, we faced challenges managing project costs due to scope creep, resulting in inaccurate estimations and budget overruns. Realizing the need for a structured approach to cost management, we implemented a CBS that categorized costs based on various components, such as the training facility, hardware, software licenses, labor, student mobilization, and training.

By utilizing the CBS, we were able to track scope creep and costs more effectively and identify areas where cost-saving measures could be implemented. For instance, we analyzed the CBS and determined that certain costs could be deducted from tax. This insight allowed us to optimize our agreements, leading to significant cost savings without compromising project deliverables.

My project management experiences taught me that utilizing breakdown structures beyond the WBS, such as the EBS, SBS, and CBS, is crucial for effective project planning and execution. These structures provide valuable insights, facilitate efficient schedule and stakeholder management, and ensure accurate cost estimation and tracking. By incorporating these lessons, project managers can enhance their ability to deliver successful projects.

NOTE: Please take note that cost control is NOT accounting.

To keep track of project costs the way accounts do it, actuals minus budget = forecast, is wrong. I'm not saying that it is not important. Our stakeholders who approved the monies/budget for our project want to know what is left.

The accountants want to know how much money must be spent next month and the rest of the year. I go into more depth about this in my other books on cost and budget control.

When we monitor and control project costs, we must determine the cost without utilizing the contingency and reserves. I have seen many projects that use all those funds to complete them, and most of the time, the project manager uses the monies for gold-plating the project. Keeping the client happy is one thing, but we should manage projects with some form of ethics.

CHAPTER 4

Other Project Structures

Resource Breakdown Structure (RBS)

The Resource Breakdown Structure (RBS) is a hierarchical structure that helps project managers organize and categorize resources needed for a project. It is an essential tool that can help identify the different types of resources needed to complete a project and their allocation and scheduled timing of the required resources..

The RBS is an extension of the Work Breakdown Structure (WBS) and is used to break down the project into its component parts. It is a useful tool for project managers in identifying the resources needed to complete each task and helps in determining the overall project budget.

The RBS is used to identify the different types of resources needed for a project, such as personnel, equipment, materials, and facilities. It is also used to identify the different levels of resources needed, such as labor, equipment, and supplies.

The RBS is created by starting with the project deliverables and breaking them down into work packages. Each work package is then broken down into its component parts, which are the resources needed to complete the package. These resources are then further broken down into their component parts until the entire project is broken down into its smallest components.

The RBS can be used to create a resource allocation matrix, which is a tool that helps project managers allocate resources to specific tasks.

The matrix identifies which resources are available, which tasks require those resources, and how much effort of each is needed.

The RBS is an essential tool for project managers in identifying the resources needed for a project. It helps in determining the overall project budget, identifying the different types of resources needed, and allocating resources to specific tasks. By using the RBS, project managers can ensure that their projects are completed on time and within budget.

NOTE: Typically used in big corporations to indicate the resources in a business unit, department or function depending on workload. A resourcing profile could also be used to show the resources required to deliver certain work that is planned in the organizations that shows current and future workload. This is normally performed via a software tool that can assist in managing the workload of resources.

Another example is when companies are executing a complete re-structuring exercise for their global or specific area-based organizations. I have been part of a few restructuring exercises where the organization is represented on all continents, has offices in most countries on that continent, and has more than one office or operation in one country. This could get quite cumbersome and very big quickly. Remember, you need resources to perform the operations and project tasks at ground level. See examples shown later.

Normally, a resource breakdown structure would be aligned with the organizational structure of a project. However, you should keep a register of people on the project and who they report to. Here is a sample of a typical project list. The resource table or list can take many formats, but you should have a list of resources for your project available from a cost-tracking and control point of view. Use these resource tables to track costs and hours. Project contracts come in various formats, and it would be wise to track the following by required period (day, week, month, quarter, etc.) *Planned*, *Actual*, and *Forecast* of each as a minimum:

- Costs,
- Requirements (how many do we need),
- Hours required.

Let's use two examples:

- Any gear (resources) required to deliver a project.
    - o Cranes, Trucks, Telehandlers,
    - o Site LDV, Mini-Buses, Cars, Cameras, Lights,
    - o Equipment & materials supply,
    - o Housing, B&B Accommodation, etc.
- People (resources) that will be working on your project
    - o Track Costs
    - o Track Man-hours
    - o Track the number of people required

This list can get extensive, but tracking what you require depends on the control you need for your project. All this information should be part of your estimate.

NOTE: Remember, developing an estimate is a combined effort of the whole team. Every discipline will supply input into the estimate and the schedule. Therefore, the disciplines should take ownership of the estimate that they provided.

Resource Category	Month 1	Month 2	Month 3
People			
- Architects	2	2	2
- Construction Workers	10	15	12
- Electricians	2	3	2
- Plumbers	2	3	2
- Painters	1	2	2
Equipment			
- Excavator	1	1	1
- Cement Mixer	1	1	1
- Crane	1	1	1
- Scaffolding	1	1	1
Materials			
- Bricks	10,000	15,000	12,000
- Cement Bags	500	800	600
- Lumber	15,000	18,000	16,000
- Roofing Shingles	2,000	2,500	2,000
- Plumbing Fixtures	50	60	50
- Electrical Wiring	2,000	2,500	2,000
Etc.			

Table 9 - Resources Requirements List

Risk Breakdown Structure (RBS)

Risk management is an essential part of project management, and it involves identifying, assessing, and addressing potential risks that can affect the outcomes of a project. To effectively manage risks, project teams use a Risk Breakdown Structure (RBS) to categorize and organize risks according to their sources, causes, and potential impacts.

An RBS is a hierarchical chart that breaks down risks into smaller and more manageable components. It is similar to a WBS but focuses on risks instead of project deliverables. The RBS provides a framework for identifying, assessing, and managing risks and helps project teams to prioritize their efforts and resources.

The RBS typically consists of three levels: Level 1, Level 2, and Level 3. Level 1 represents the highest level of categorization, while Level 3 represents the lowest level of detail.

Level 1 of the RBS is usually based on the sources of risks, such as internal, external, or project-specific risks. Level 2 categorizes risks according to their causes, such as technical, organizational, or environmental risks. Level 3 breaks down risks into specific events or conditions that could occur, such as weather conditions, equipment failure, or human error.

To create an RBS, project teams should start by identifying potential risks and their sources. They should then break down these risks into smaller components and categorize them according to their causes and potential impacts. This process should involve all stakeholders, including the project team, clients, suppliers, and other relevant parties.

Once the RBS is complete, project teams can use it to assess the severity of each risk, prioritize their efforts and resources, and develop appropriate risk management strategies. The RBS can also help project teams to communicate and report on risks to stakeholders, monitor risks throughout the project lifecycle, and make informed decisions based on risk analysis.

In conclusion, the Risk Breakdown Structure (RBS) is a powerful tool for managing project risks. It helps project teams categorize and organize risks according to their sources, causes, and potential impacts and provides a framework for identifying, assessing, and managing risks. By using an RBS, project teams can prioritize their efforts and resources, develop appropriate risk management strategies, and make informed decisions based on risk analysis.

NOTE: A Risk breakdown structure (RBS) would typically be used when delivering a project over a wide geographical area, and the risks pertaining to that graphical area could impact the project overall. Keeping track of the risks in an RBS could assist in highlighting the areas of concern very early.

Quality Breakdown Structure (EBS)

The Quality Breakdown Structure (QBS), also known as the Evaluation Breakdown Structure (EBS), is a hierarchical representation of the factors that contribute to the overall quality of a product or project. It is a tool used in project management to identify, assess, and prioritize the key quality aspects that need to be addressed.

The QBS provides a systematic approach to quality management by breaking down the concept of quality into smaller, manageable components. These components represent different dimensions of quality that can be evaluated and measured. By analyzing each component, project managers can gain a comprehensive understanding of the quality requirements and identify areas that need improvement.

The structure of the QBS is typically developed through a collaborative effort involving stakeholders, project team members, and subject matter experts. It starts with identifying the main quality objectives, which are divided into smaller, more specific quality factors. These factors can be further broken down into sub-factors, creating a hierarchical structure that represents the different levels of detail.

82

Each level of the QBS contains quality indicators or metrics that are used to evaluate the performance of the corresponding quality factor. These indicators can be quantitative or qualitative, depending on the nature of the quality aspect being assessed. For example, in a software development project, the QBS may include indicators such as code maintainability, user interface usability, and system reliability.

One of the benefits of using the QBS is that it enables project managers to prioritize quality improvements based on the relative importance of each quality factor. By assigning weights or scores to the indicators at each level, project managers can determine the overall contribution of each factor to the final product quality. This helps in allocating resources and efforts effectively to address the critical quality aspects.

The QBS can also be used as a communication tool to convey the quality requirements to the project team and stakeholders. It provides a clear and structured overview of the quality aspects that need attention, facilitating discussions and decision-making related to quality management.

In summary, the Quality Breakdown Structure (QBS) is a hierarchical representation of the quality factors that need to be evaluated and managed in a project or product. It helps identify, assess, and prioritize the key quality aspects, enabling project managers to allocate resources effectively and improve the overall quality.

NOTE: A quality breakdown structure (QBS) would typically be used when you have a very stringent quality system to follow. For instance, when constructing or maintaining a facility, such as a nuclear plant, or as simple as manufacturing a bicycle or manufacturing some kind of widget.

Stakeholder breakdown structure (SBS)

The Stakeholder Breakdown Structure (SBS) is a project management tool that helps identify and categorize stakeholders involved in a project. It visually represents the project's stakeholders and their relationships, allowing project managers to engage and manage the stakeholders effectively. You could include internal and external stakeholders on the same WBS or split them for ease of use.

The external stakeholders should sometimes see how they influence other external stakeholders with their decisions. If the various stakeholders belong to various political parties, you must make sure that your stakeholder management plan covers conflict issues that might occur on your project. These types of conflict or disagreement issues can add years to your project.

The SBS is similar to a Work Breakdown Structure but focuses solely on stakeholders. It breaks down the project's stakeholders into smaller, more manageable groups, allowing for a more targeted approach in understanding their needs, expectations, and influence on the project.

To create an SBS, project managers typically follow these steps:

1. **Identify all stakeholders:** The first step is to identify all individuals, groups, or organizations who have an interest or influence in the project. This includes project sponsors, team members, customers, suppliers, regulatory bodies, and other relevant parties.

2. **Categorize stakeholders:** Once all stakeholders are identified, they are classified into different categories based on their relationships with the project. Common categories include internal stakeholders (such as project team members and executives), external stakeholders (such as customers and suppliers), and

regulatory stakeholders (such as government agencies and industry associations).

3. **Determine stakeholder attributes:** For each stakeholder, it is important to identify their specific attributes, such as their roles, responsibilities, expectations, and level of influence. This helps in understanding the stakeholder's needs and how they may impact the project.

4. **Define stakeholder relationships:** Stakeholders are typically interconnected and have varying degrees of influence on the project. Mapping out these relationships helps project managers understand the dynamics between stakeholders and their potential impact on project outcomes. This can be done using techniques like stakeholder mapping or influence diagrams.

5. **Analyze stakeholder engagement strategies:** With a clear understanding of stakeholder attributes and relationships, project managers can develop appropriate strategies to engage and manage stakeholders effectively. This may include regular communication, stakeholder meetings, feedback sessions, or other tailored approaches to address their specific needs and concerns.

The Stakeholder Breakdown Structure (SBS) provides several benefits for project management. It helps project managers:

- Identify and understand the various stakeholders involved in the project.

- Prioritize stakeholder engagement efforts based on their significance and influence.

- Determine potential conflicts or areas of alignment between stakeholders.

- Develop appropriate communication and engagement strategies to ensure stakeholder satisfaction.

- Minimize risks associated with stakeholder dissatisfaction or non-alignment with project goals.

In summary, the Stakeholder Breakdown Structure is a valuable tool for project managers to systematically categorize and analyze stakeholders involved in a project. By understanding their attributes, relationships, and engagement strategies, project managers can effectively manage stakeholder expectations and ensure project success.

Communication Breakdown Structure (CBS)

The Communication Breakdown Structure (CBS) is a project management tool that helps identify, organise, and document the communication activities within a project. It provides a hierarchical breakdown of project stakeholders' communication requirements, responsibilities, and relationships.

The CBS is derived from the Work Breakdown Structure (WBS), which defines the project scope and deliverables. It outlines the communication needs and channels required to complete each work package or deliverable. Breaking down communication activities ensures that all relevant stakeholders are included in the communication process and helps avoid any misunderstandings or gaps in communication.

At its core, the CBS consists of three main components:

1. **Communication Categories:** These are the major groups or categories of communication activities within the project. Examples may include internal communication, external communication, stakeholder engagement, or team coordination. Each category represents a broad area of communication needs.

2. **Communication Elements:** These are the specific communication activities within each category. It includes tasks such as meetings,

reports, emails, presentations, or status updates. Each element represents a specific communication requirement or deliverable.

3. **Communication Relationships:** This component defines the relationships and dependencies among different communication elements and categories. It helps to identify which communication activities are dependent on others and how they interact with each other.

The CBS helps project managers to:

1. **Identify Communication Requirements:** By breaking down the project scope into communication categories and elements, the CBS helps in identifying all the necessary communication activities needed for successful project execution. This ensures that no important communication aspect is overlooked.

2. **Assign Responsibilities:** The CBS helps in clearly defining the responsible parties for each communication element. It ensures that each stakeholder knows their role and responsibilities in communication.

3. **Plan Communication Activities:** By organizing communication activities in a hierarchical structure, the CBS allows project managers to plan the timing, frequency, and channels for each communication element. It helps in creating a comprehensive communication plan for the project.

4. **Track and Control Communication:** The CBS serves as a reference document for tracking and controlling communication activities. It enables project managers to monitor the progress of communication tasks and approval cycles, identify bottlenecks or delays, and take corrective actions if necessary.

In conclusion, the Communication Breakdown Structure (CBS) is a valuable tool for project managers to manage and control communication within a project effectively. It helps identify, organise, and document the communication requirements, responsibilities, and relationships among project stakeholders, ensuring effective communication throughout the project lifecycle.

NOTE: Managing a large portfolio of projects spanning various business units, local area-based companies with an overall Head Office (HO) that must approve projects can get very complicated. Here is an example of a communications structure that will keep everyone informed about the projects submitted to which approval body. The sample below is a small extract of projects submitted for approval.

A Communication Breakdown Structure (CBS) is required when issues such as project approvals (project value-defined levels of approval) are complicated and timing or schedule clashes are unavoidable. Keeping your stakeholders informed of what projects are submitted for approval can be demonstrated with a straightforward breakdown structure to keep the project teams aligned.

A reminder to Keep it Simple, Simon:

The KISS principle.

The communication structure below reflects a complicated CBS setup. It must be noted that many different structures can be set up to depict a CBS.

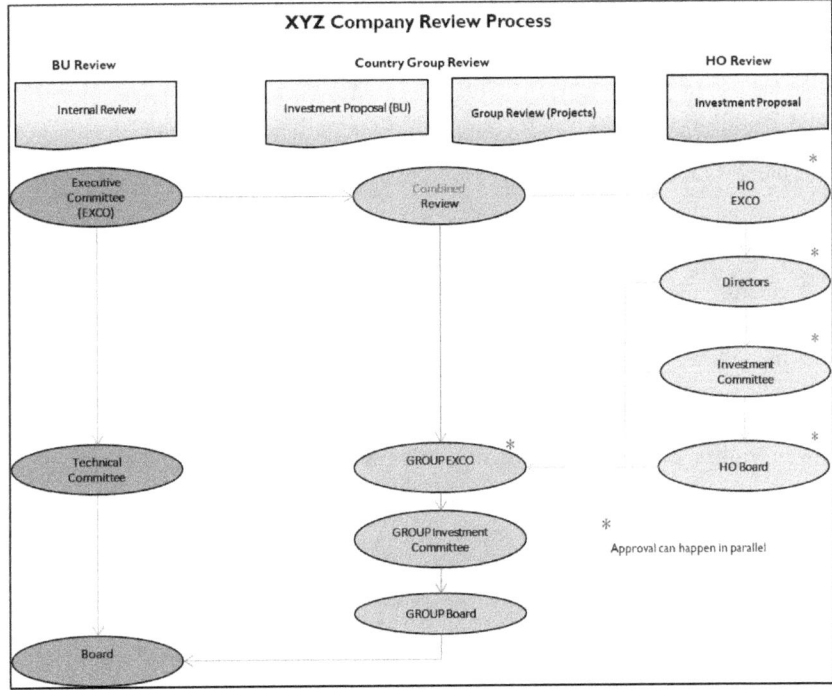

Figure 12 - A complicated Communications structure for Approvals

#	Project No	Project Type	STAGE	Category	Area	RANKING	FUNCTION	Description	Auth Year
1	Project 294	Expansion	EXECUTION	Small	Mine 3	12	PLANT	Project 294	2022
2	Project 59	Maintenance	CONCEPT	Medium	Mine 4	12	EXPLORATION	Project 59	2022
3	Project 67	Maintenance	CONCEPT	Small	Mine 4	11	EXPLORATION	Project 67	2023
4	Project 181	Expansion	EXECUTION	Medium	Mine 2	16	PLANT	Project 181	2022
5	Project 170	Expansion	EXECUTION	Small	Mine 4	8	INFRASTRUCTURE	Project 170	2022
6	Project 77	Expansion	CONCEPT	Medium	Mine 5	12	EXPLORATION	Project 77	2022
7	Project 36	Expansion	EXECUTION	Medium	Mine 1	16	PLANT	Project 36	2023
8	Project 140	Maintenance	FEASIBILITY	Medium	Mine 1	16	MINE	Project 140	2022
9	Project 135	Expansion	FEASIBILITY	Medium	Mine 4	18	MINE	Project 135	2022
10	Project 136	Expansion	EXECUTION	Medium	Mine 2	8	INFRASTRUCTURE	Project 136	2023
11	Project 137	Expansion	EXECUTION	Medium	Mine 1	16	MINE	Project 137	2023
12	Project 138	Expansion	EXECUTION	Medium	Mine 1	16	MINE	Project 138	2023
13	Project 61	Expansion	EXECUTION	Medium	Mine 4	18	MINE	Project 61	2024
14	Project 62	Expansion	EXECUTION	Small	Mine 3	8	INFRASTRUCTURE	Project 62	2024

Table 10 - List of Projects for Approval

Key Tips: Other Breakdown Structures

As a seasoned project management professional, I understand the importance of using other breakdown structures in addition to the traditional work breakdown structure (WBS) to ensure the successful delivery of any project. These additional breakdown structures help me better organise and manage a project's various aspects.

Here are some key tips I have learned about the need for other breakdown structures:

On the other hand, the Resource breakdown structure (RBS) allows for a comprehensive understanding of the human, physical, and material resources required for a project. Breaking down the project into its various components makes identifying the specific resources needed for each task easier. This structure aids in resource allocation, capacity planning, and ensures that the right resources are available at the right time. Instead of using a RBS, a straight forward list can be used to manage a project's resources, costs, and hours.

When it comes to managing risks, the *Risk breakdown structure (RBS)* is an invaluable tool. By identifying and categorizing potential risks, this structure allows for effective risk mitigation strategies to be developed and implemented. It helps prioritise risks, allocate resources to manage them, and ensure the project team is prepared for unforeseen circumstances.

The *Quality breakdown structure (QBS)* plays a critical role in ensuring that the project meets or exceeds the expected standards. Breaking down quality requirements into specific components makes monitoring and controlling the quality of deliverables throughout the project lifecycle much easier. This structure helps set quality standards, conduct inspections, and implement quality assurance measures.

In any project, stakeholders play a significant role. The *Stakeholder breakdown structure (SBS)* helps identify and categorize stakeholders based on their level of influence, interest, or involvement in the project. This structure aids in stakeholder management, communication planning, and ensures that the project team is aware of the various stakeholders and their specific needs or requirements.

Lastly, effective communication is vital for the success of any project, and the *Communication breakdown structure (CBS)* facilitates this. By breaking down the project communication needs, channels, and methods, this structure ensures that the right information is shared with the right stakeholders at the right time. It helps develop a communication plan, establish communication protocols, and maintain transparency throughout the project.

In conclusion, using other breakdown structures in addition to the WBS is crucial for successful project management. These structures provide a systematic approach to managing various aspects of a project, ensuring better control, communication, and, ultimately, the successful delivery of the project. As a project manager, I highly recommend incorporating these breakdown structures into your project management toolkit for optimal results. By incorporating these additional breakdown structures into my project during planning and execution, I am able to enhance project outcomes and ensure its overall success.

NOTE: Not all of the breakdown structures have to be used, and the project team must utilize these structures to better manage their own areas of control.

Lessons Learnt: Other Breakdown Structures

Lesson 1: Importance of Breakdown Structures

Throughout my project management journey, I have learned the significance of using breakdown structures other than the Work Breakdown Structure (WBS). While the WBS provides a comprehensive breakdown of project deliverables, it is equally essential to explore other breakdown structures to ensure effective project planning and execution.

One such breakdown structure I have encountered is the Organizational Breakdown Structure (OBS). This structure helps identify and assign project responsibilities to various organizational individuals or departments. By clearly defining roles and responsibilities, the OBS facilitates smooth communication and coordination among team members, leading to improved project outcomes.

NOTE: This breakdown is key as it assists with the allocation of resources on a project, which in turn assists with the estimating, budgeting, and resources required on a project.

For instance, in a recent project, we encountered a critical issue related to stakeholder management. Despite having a well-defined WBS, we struggled with ineffective communication and a lack of accountability. This is an important lesson, as normally, in projects, some leaders tend to shun their responsibilities when things go wrong. If a Stakeholder Breakdown Structure (SBS) is in place, then this is a non-issue. Realizing the need for an additional breakdown structure, we implemented the OBS, which allowed us to identify the key stakeholders involved and assign specific responsibilities to each team member. This helped us streamline communication channels and mitigate potential conflicts, resulting in improved stakeholder management throughout the project lifecycle.

Lesson 2: Benefits of a Risk Breakdown Structure (RBS)

In my project management experience, I have learned the importance of utilizing the Risk Breakdown Structure (RBS) as an essential tool for effective risk management. The RBS provides a systematic approach to identifying, analyzing, and mitigating project risks, ensuring proactive risk management throughout the project.

In a recent construction project, we faced numerous unforeseen delays due to inclement weather conditions. These delays had a significant impact on the project timeline and budget. Realizing the need to implement a comprehensive risk management approach, we developed an RBS that categorized risks based on their causes, such as weather, resource availability, and technical issues.

By utilizing the RBS, we identified potential risks early on and developed appropriate contingency plans. For instance, in the case of adverse weather conditions, we incorporated weather monitoring systems and contingency plans to mitigate the impact on project timelines. This proactive approach helped us minimize delays and ensure successful project completion.

Lesson 3: Value of a Cost Breakdown Structure (CBS)

In my project management journey, I have discovered the significance of utilizing the Cost Breakdown Structure (CBS) to manage project costs effectively. The CBS provides a hierarchical breakdown of project costs, allowing for accurate estimation, budgeting, and tracking of expenses throughout the project lifecycle.

In a recent software development project, we faced challenges in managing project costs due to inaccurate estimations and budget overruns. Realizing the need for a structured approach to cost management, we implemented a

CBS that categorized costs based on various components, such as hardware, software licenses, labor, and training.

By utilizing the CBS, we were able to track costs more effectively and identify areas where cost-saving measures could be implemented. For instance, by analyzing the CBS, we identified that certain software licenses were underutilized, resulting in unnecessary expenses. This insight allowed us to optimize our software licensing agreements, leading to significant cost savings without compromising project deliverables.

My project management experiences taught me that utilizing breakdown structures beyond the WBS, such as the OBS, RBS, and CBS, is crucial for effective project planning and execution. These structures provide valuable insights, enable proactive risk management, facilitate efficient stakeholder management, and ensure accurate cost estimation and tracking. By incorporating these lessons, project managers can enhance their ability to deliver successful projects.

CHAPTER 5

Techniques for Breaking Down Projects

Responsibility Assignment Matrix (RAM):

The RAM is a tool for demonstrating the relationship between work packages or activities and project team members. Its primary goal is to clarify who is in charge of what within the project. The RACI or RASCI chart is the most common type of RAM, and it divides project responsibilities into four roles: Responsible, Accountable, Supported, Consulted, and Informed. The RACI or RASCI does not specifically fall into a breakdown structure, but can be used in a table format that depicts individuals' main Responsibility and Accountability on a project. Whereas the RAM ensures that duties and responsibilities are clearly defined, making it clear who is in charge of execution, who has decision-making authority, who needs to be consulted, and who should be kept up to date on the status of various tasks.

Here are two RACI Charts for an organization that goes through various project life-cycle phases. In this case, they would reflect a RACI for each phase, Concept, Pre-Feasibility, Feasibility, and Execution. These two depict a project's concept and feasibility stage or phase and show all the departments and disciplines that must be consulted during that phase.

NOTE: A RACI is usually depicted in a table that demonstrates the matrix, not a diagram or a type of Matrix or Breakdown Structure.

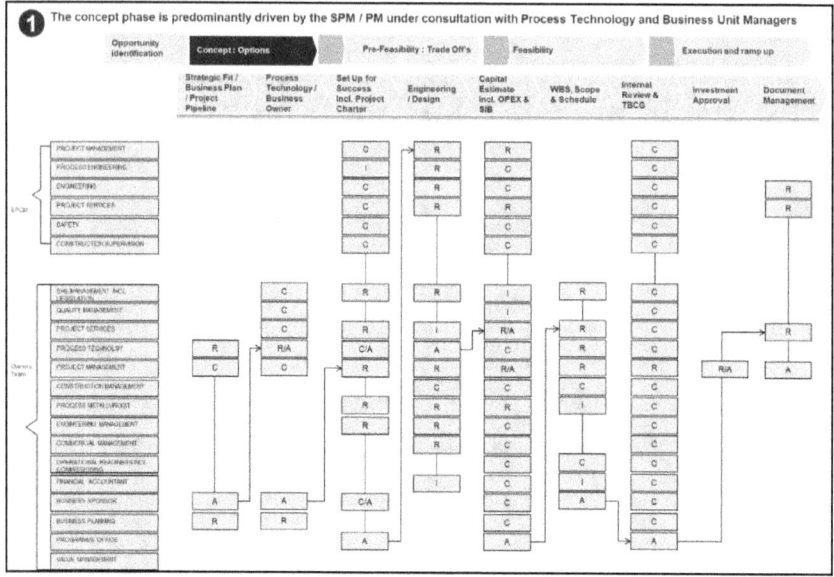

Figure 13 - RACI chart -Concept Phase

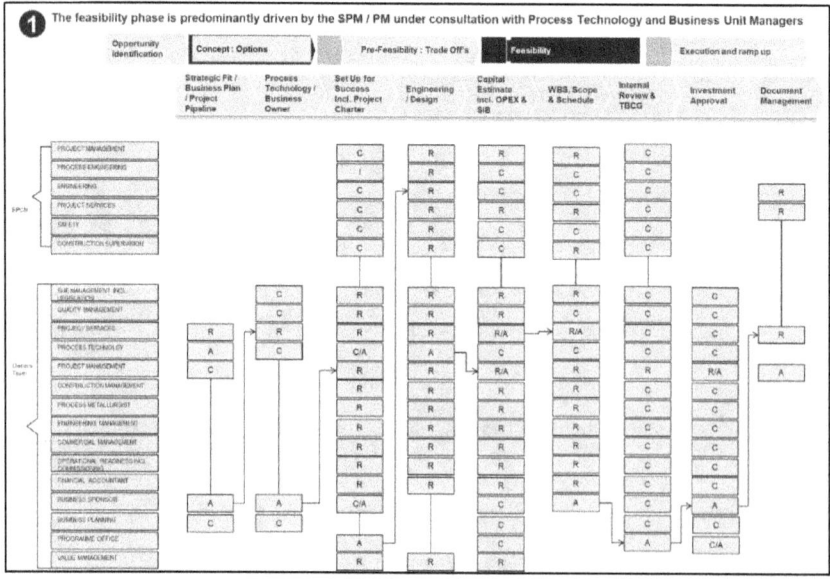

Figure 14 - RACI chart – Feasibility Phase

Network Diagrams

Network diagrams are essential for any project team member involved in developing a project management work breakdown structure (WBS) and other structures. They visually represent the project's schedule, showing the sequence of activities and their dependencies. Network diagrams are also known as flowcharts or PERT charts (Program Evaluation and Review Technique).

Network diagrams can help project team members to identify the critical path, which is the sequence of activities that must be completed on time to meet the project's deadline. They can also help project team members determine the project's duration and the resources needed to complete it.

To create a network diagram, project team members should follow these steps:

1. Identify the activities – List all the activities needed to complete the project. Each activity should have a unique identifier, a name, and a duration.

2. Determine the dependencies – Identify the relationships between the activities. Some activities must be completed before others can start. These dependencies can be categorized as finish-to-start, start-to-start, finish-to-finish, or start-to-finish.

3. Draw the diagram – Use arrows to connect the activities, showing the dependencies between them. The arrows should point from the predecessor activity to the successor activity.

4. Add the duration – Add the duration of each activity to the diagram. This will show the estimated time required to complete each activity.

5. Identify the critical path – Identify the sequence of activities that must be completed on time to meet the project's deadline. This is the critical path, and it should be highlighted on the diagram.

6. Review and update the diagram – The network diagram should be reviewed and updated regularly to reflect any changes in the project schedule.

In conclusion, network diagrams are a powerful tool for project team members involved in project management work breakdown structure (WBS) and other structures. They help to visualize the project schedule, identify the critical path, and determine the resources needed to complete the project. By following the steps outlined above, project team members can create an accurate and effective network diagram.

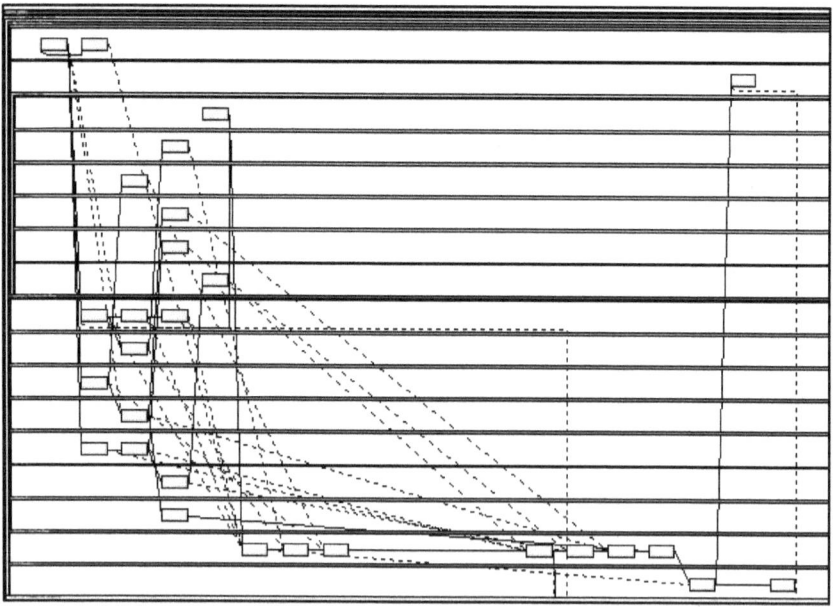

Figure 15 - Typical Network Diagram

Gantt Charts

A Gantt chart is a graphical representation of a project schedule. It is commonly used in project management to show the timing of tasks and the relationships between them. A Gantt chart is a useful tool for project team members because it provides a clear and concise picture of the project schedule. The schedule bar chart (Gantt chart), usually reflects the WBS and can depict the WBS in various levels, such as Level 1, Level 3, and the like.

This chart is named after Henry Gantt, an American engineer and management consultant who developed the technique in the early 20th century. Gantt charts are now widely used in project management and are an essential tool for project team members.

A Gantt chart is a bar chart showing project activities' start and end dates. Each activity is represented by a horizontal bar, and the bar's length represents the activity's duration. The bars are arranged in a time sequence, so the chart shows the order in which the activities will be performed.

The Gantt chart also shows the dependencies between activities. Dependencies are the relationships between activities that determine the order in which they must be performed. For example, Activity A may need to be completed before Activity B can start. The Gantt chart shows this relationship by placing Activity A before Activity B on the chart.

The Gantt chart is an excellent tool for project team members because it provides several benefits. First, it provides a clear and concise picture of the project schedule. This helps team members understand the timing of tasks and the dependencies between them. Second, it helps team members identify potential problems and bottlenecks in the project schedule. For example, if two activities are scheduled to start at the same time, the Gantt chart will show this as a conflict.

Finally, the Gantt chart is an essential tool for project team members because it provides a basis for communication and coordination. By using the Gantt chart, team members can discuss the project schedule and identify potential issues. This creates a more collaborative and effective project team.

In conclusion, the Gantt chart is a valuable tool for project team members. It provides a clear and concise picture of the project schedule, helps identify potential problems, and promotes communication and coordination among team members. Using the Gantt chart allows project team members to work more effectively and efficiently, leading to a more successful project outcome.

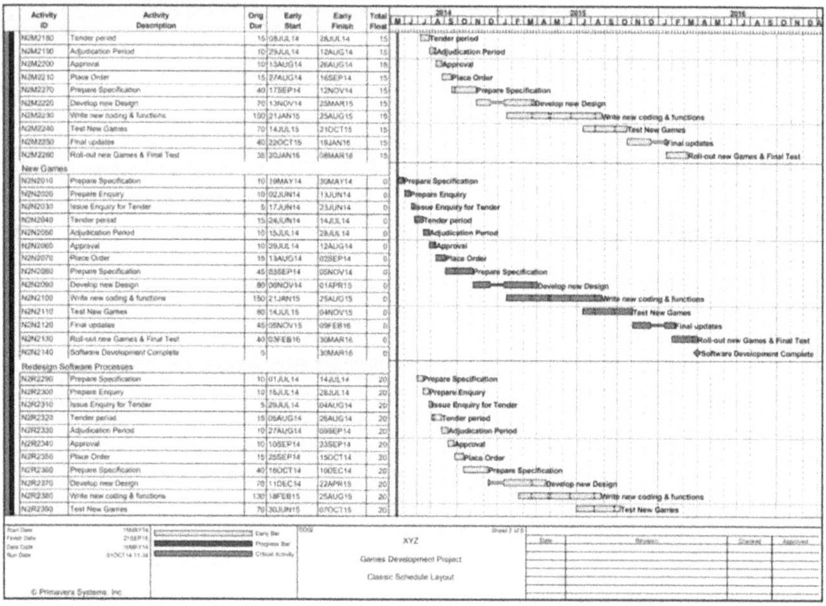

Figure 16 - Typical Gantt Chart and CPA

Critical Path Analysis (CPA)

Critical Path Analysis (CPA) is a project management technique used to identify the most critical tasks in a project and determine the shortest possible time to complete the project. The critical path is the sequence of tasks that must be completed on time to ensure the project is completed within the defined timeline.

It is important to reflect your project WBS in the schedule. By using the CPA, it will very clearly indicate which WBS elements are on the critical path.

CPA involves identifying all the tasks in a project, their dependencies, and the estimated time required to complete them. The critical path is then determined by identifying the longest sequence of tasks that must be completed without delay.

To carry out a CPA, project team members must first create a project schedule that includes all the tasks and their dependencies. This can be done using a project management software tool or by creating a detailed Gantt chart.

Once the project schedule is in place, the team can identify the critical path by analyzing the duration of each task and its dependencies. The critical path is the sequence of tasks that has the longest duration and cannot be delayed without delaying the entire project.

The critical path analysis allows project team members to identify the most important tasks in a project and prioritize their efforts accordingly. It also helps to identify potential delays and bottlenecks that could impact the project timeline.

By using CPA, project team members can ensure that they focus their efforts on the most critical tasks, and allocate resources effectively to ensure the project is completed on time. It also helps to identify potential risks and challenges that may arise during the project and develop contingency plans to mitigate them.

In conclusion, Critical Path Analysis is an essential project management tool that helps project team members identify the most critical tasks in a project, prioritize their efforts, and ensure the project is completed within the defined timeline. By using CPA, project team members can effectively manage resources, identify potential risks, and develop contingency plans to ensure project success.

Mind Mapping

One of the most effective tools for breaking down projects is mind mapping. Mind mapping is a graphical technique that allows you to visually represent ideas and concepts. It can help you organize your thoughts and ideas, and is an excellent way to brainstorm new ideas and solutions.

When you start a new project, trying to organise all the tasks and details can be overwhelming. This is where mind mapping can help. Creating a mind map allows you to break down the project into smaller, more manageable pieces. This will make it easier to see the big picture and identify any potential issues.

To create a mind map, start by writing the main idea or concept in the center of the page. From there, branch out and add subtopics, tasks, and details. Use colors, symbols, and images to help visually distinguish different ideas and concepts. You can use software like MindJet or Mindomo to create digital mind maps, or you can use pen and paper.

Mind mapping is a great way to involve your team in the project planning process. You can use it in brainstorming sessions to generate new ideas, or you can use it to assign tasks and responsibilities. By involving your team in the mind-mapping process, you can ensure that everyone is on the same page and that everyone understands their roles and responsibilities.

Mind mapping is also a great way to track progress and make adjustments as needed. As you complete tasks and move through the project, you can update your mind map to reflect any changes or adjustments. This will help you stay organized and ensure that you are on track to meet your goals and deadlines.

In conclusion, mind mapping is an effective tool for breaking down projects and organizing your thoughts and ideas. It can help you visualize the big picture and identify any potential issues or challenges. By involving your team in the mind-mapping process, you can ensure that everyone is on the same page and that everyone understands their roles and responsibilities. Use mind mapping to track progress and make adjustments as needed, and you will be well on your way to successfully completing your project.

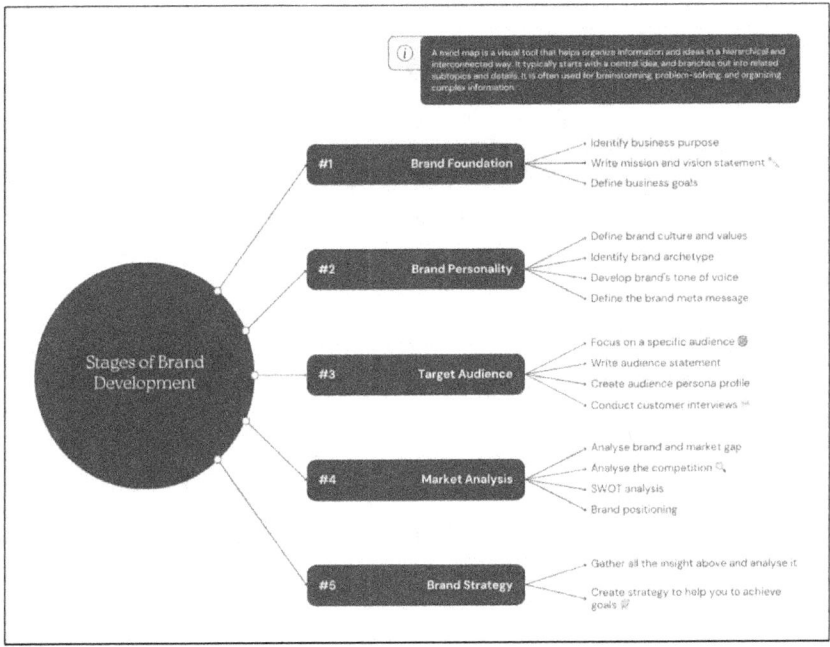

Figure 17 - Typical Mind Mapping Diagram

Affinity Diagrams

Affinity diagrams are a valuable tool for project teams to organize and make sense of large amounts of information. These diagrams are commonly used in brainstorming sessions to group and categorize ideas, issues, and concerns related to a project.

To create an affinity diagram, project team members start by generating a list of ideas or issues related to the project. Each idea is written on a separate sticky note or index card. Once all ideas have been recorded, team members begin to group similar ideas together based on common themes or patterns. These groups become the categories or clusters of the affinity diagram.

By organizing ideas in this way, project teams can identify key themes and issues that must be addressed to move the project forward. Affinity diagrams can also be useful in identifying areas of overlap or redundancy in the project, allowing teams to streamline their efforts and focus on the most critical tasks.

One of the benefits of affinity diagrams is that they allow for a high degree of collaboration among project team members. By working together to group and categorize ideas, team members gain a shared understanding of the project and its goals. This can lead to stronger communication and more effective decision-making throughout the project lifecycle.

Another advantage of affinity diagrams is that they are easily customizable to fit the needs of any project. Larger projects may require multiple diagrams to capture all of the ideas and issues related to the project, while smaller projects may only need a single diagram. Additionally, the categories or clusters within an affinity diagram can be adjusted as needed to reflect changes in the project or new information that emerges over time.

Affinity diagrams are a powerful tool for project teams organising and making sense of large amounts of information. By working collaboratively to group and categorize ideas, project teams can gain a shared understanding of the project and its goals, leading to more effective decision-making and a stronger overall outcome.

Fishbone Diagrams

Fishbone diagrams, also known as Ishikawa diagrams or cause-and-effect diagrams, are a powerful tool for identifying and analyzing the root causes of problems in a project. They are particularly useful in situations where the cause of a problem is not immediately obvious, or where multiple factors may be contributing to the issue.

104

The basic structure of a fishbone diagram is simple: a central spine represents the problem you are trying to solve, and branches extending from the spine represent different possible causes or contributing factors. Each branch can be further subdivided into more specific causes, creating a branching structure resembling a fish's skeleton (hence the name).

To create a fishbone diagram, begin by identifying the problem you want to analyze. This could be anything from a missed deadline to a quality issue in a product or service. Once you have identified the problem, brainstorm different possible causes and write them down as branches extending from the spine.

Next, consider each potential cause in more detail and try to identify any sub-causes or contributing factors that may be relevant. For example, suppose one of the potential causes of a quality issue is poor communication between team members. In that case, you might identify sub-causes such as language barriers, lack of clear instructions or feedback, or a lack of documentation.

Continue this process until you have identified all relevant causes and sub-causes and created a comprehensive diagram that captures the full range of factors contributing to the problem. You can then use this diagram to prioritize and address the most important causes and develop targeted solutions that address each one.

Fishbone diagrams are a useful tool for project team members because they encourage a systematic and thorough approach to problem-solving. They also help teams identify underlying issues contributing to multiple problems and develop more effective solutions that address the root cause rather than just the symptoms.

By incorporating fishbone diagrams into your project management toolkit, you can increase your team's ability to analyze and solve complex problems and improve your projects' overall quality and success.

The fishbone diagram, also known as an Ishikawa diagram or a cause-and-effect diagram, is typically used for the following purposes:

- **Root Cause Analysis:** Fishbone diagrams are primarily used for identifying and analyzing the root causes of a problem or an issue. By visually mapping out the various potential causes and their relationships, teams can pinpoint the underlying reasons for a particular problem.
- **Problem Solving:** They are valuable in problem-solving processes, helping teams systematically investigate and address issues. By understanding the causes, teams can develop effective solutions to mitigate or eliminate the problem.
- **Quality Improvement:** Fishbone diagrams are commonly employed in quality improvement initiatives such as Six Sigma and Total Quality Management (TQM). They help organizations identify quality-related issues and implement corrective actions.
- **Process Improvement:** Teams can use fishbone diagrams to analyze and improve organizational processes. It allows for a structured approach to process optimization by identifying factors contributing to inefficiencies or defects.
- **Risk Assessment:** Fishbone diagrams can be used to assess and manage risks. By identifying potential root causes of risks, organizations can take preventive measures to minimize the likelihood of adverse events.
- **Product Development:** In product development, fishbone diagrams can help teams identify factors affecting product quality or delays in development. This aids in delivering products that meet customer expectations.
- **Project Management:** Project managers may use fishbone diagrams to identify and address project-related issues and delays. It helps in understanding the causes of project setbacks and finding ways to keep the project on track.
- **Brainstorming:** Fishbone diagrams are also useful in brainstorming sessions. They provide a structured framework for teams to explore and generate ideas about potential causes of a

problem or challenges.

- **Cross-Functional Collaboration:** They promote collaboration among cross-functional teams because they visually represent the problem and its causes, making it easier for different team members to understand and contribute to the analysis.

As is demonstrated, fishbone diagrams are a versatile tool that helps organizations and teams investigate, analyze, and address a wide range of issues, making them an essential part of problem-solving and continuous improvement efforts. Most of the items listed above can assist with a breakdown structure development.

Decision Trees

Decision trees are an effective tool for project team members to use in making decisions. A decision tree is a graphical representation of a decision-making process. It is a tree-like model of decisions and their possible consequences, including chance event outcomes, resource costs, and utility. Decision trees help project team members identify the best course of action given the available options and likely outcomes.

To create a decision tree, project team members must begin by defining the decision they are facing. They then identify the available options and the potential outcomes of each option. Factors that can impact the success of an option should be considered, such as resource availability, time constraints, and stakeholder expectations. Project team members should also be mindful of any risks or uncertainties associated with each option.

Once all options and outcomes have been identified, project team members can begin to construct the decision tree. The tree is built by starting with the initial decision and then branching out to show the possible outcomes of each option. Each branch represents a decision point, and each node represents a possible outcome. The probabilities of each outcome should be factored into the decision tree to give a more accurate representation of the likelihood of each outcome.

Decision trees can be used in a variety of project management contexts. For example, they can be used when deciding whether to pursue a new project, when deciding which vendor to select for a project, or when deciding whether to take a particular risk. Decision trees can help project team members to make more informed decisions by allowing them to see the potential outcomes of each option.

In conclusion, decision trees are an effective tool for project team members to use in making decisions. They provide a graphical representation of a decision-making process, allowing project team members to identify the best course of action given the available options and likely outcomes. Decision trees can be used in various project management contexts and help project team members make more informed decisions.

NOTE: Both the fishbone and decision tree diagram can assist you and your team to analyze your projects structures, such as the WBS, schedule, Risks, Change Control, to name a few.

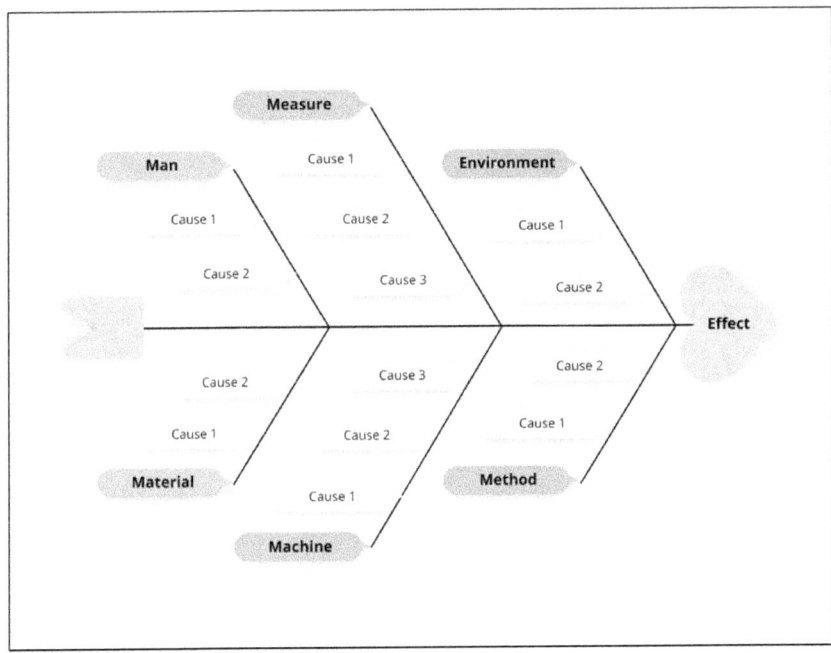

Figure 18 - Typical Fishbone Diagram

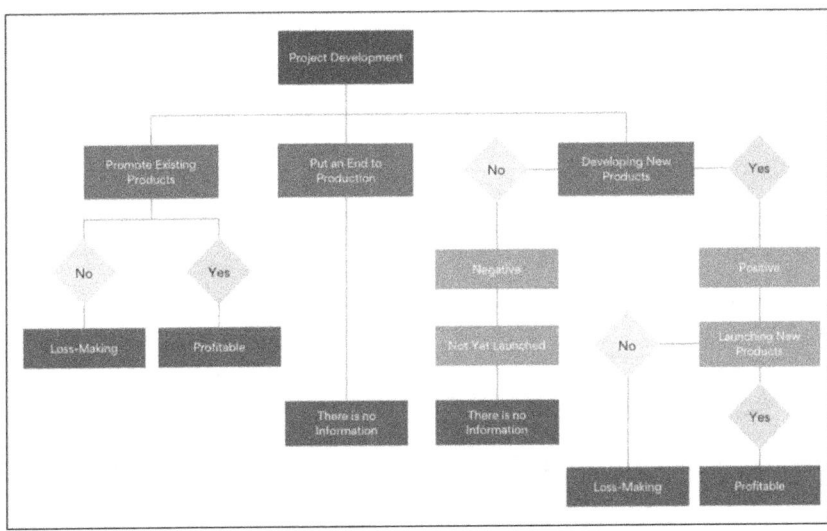

Figure 19 - Typical Decision Tree

Key Tips: Structures & Techniques

Prioritize tasks: When breaking down a project, it's important to prioritize tasks based on their importance and urgency. This helps in creating a clear roadmap for completing the project successfully.

Divide and conquer: Break down the project into smaller, more manageable tasks. This allows you to focus on one task at a time, increasing productivity and reducing overwhelm.

Use project management tools: Utilize project management software or tools to help break down the project into smaller tasks, assign deadlines, and track progress. This ensures that nothing falls through the cracks and keeps everyone on the same page.

Delegate tasks: If possible, delegate certain tasks to team members or colleagues who have the necessary skills and expertise. Delegating tasks lightens your workload and allows everyone to contribute their best to the project.

Set milestones: Break down the project into milestones or checkpoints. This helps in measuring progress and provides a sense of accomplishment as you complete each milestone.

Timeboxing: Allocate specific time blocks for working on different tasks. This technique helps in maintaining focus and prevents tasks from dragging on indefinitely.

Seek input from others: Collaborate with team members or stakeholders to get their input and perspective on breaking down the project. This can lead to fresh ideas and insights that you may have overlooked.

Use visual aids: Visualize the project by creating flowcharts, diagrams, or mind maps. This visual representation helps in understanding the project structure and identifying any potential gaps or dependencies.

Consider dependencies: Identify any tasks or activities that are dependent on others. By understanding the dependencies, you can plan accordingly and ensure that the project progresses smoothly.

Stay flexible: Understand that breaking down a project is an iterative process. As you dive deeper into the project, new information or challenges may arise, requiring adjustments to the initial breakdown. Stay flexible and adapt as needed.

Lessons Learnt: Techniques for Breaking Down Projects

Lesson learnt: Responsibility Assignment Matrix (RAM)

During my experience in construction projects in the oil and gas and mining industries, I learned the importance of creating a Responsibility Assignment Matrix (RAM) early on in the project. This matrix helped define and allocate responsibilities to different team members and stakeholders. Having a well-defined RAM made it easier to identify who was accountable for specific tasks and deliverables, improving overall project coordination and communication.

RAM helps identify who is accountable for specific tasks, which avoids confusion and ensures efficient execution. It also emphasizes the need for regular communication and collaboration among team members to ensure the smooth flow of work and timely completion of projects. Additionally, project managers should regularly review and update the RAM to accommodate any changes or adjustments in the project scope or team structure.

Lesson learnt: Network diagrams.

Working on construction projects in the oil and gas, information technology, and mining sectors taught me the significance of network diagrams. These diagrams helped visualise the sequence of activities and their dependencies, allowing for better project planning and scheduling. By creating a comprehensive network diagram, I was able to identify critical paths and areas where adjustments could be made to optimize the project timeline.

NOTE: When I started out, we printed a schedule's complete network diagram and spread it out on the office floor. Then, get the construction team to follow the correct logic for the shutdown project.

Network diagrams help in identifying the critical path, i.e., the sequence of activities that determine the project's overall duration. They also aid in understanding the interdependencies between tasks and potential bottlenecks, allowing project managers to allocate resources effectively. However, it is crucial to update network diagrams regularly to reflect any changes in project timelines or (especially) task dependencies.

Lesson learnt: Gantt Charts

In the context of construction projects in the oil and gas and mining industries, Gantt Charts proved to be an invaluable tool for project management. Lessons learnt include the importance of creating a comprehensive project schedule that includes all tasks, milestones, and deliverables. Gantt Charts help visualise the project timeline, track progress, and manage resources effectively. Using Gantt Charts, I could schedule and track project activities, set realistic deadlines, and monitor progress.

It highlights the need to regularly monitor and update the Gantt Chart to reflect any changes or deviations from the original plan. This visual representation of tasks and their durations helped me to effectively communicate project timelines to stakeholders and identify potential delays or issues that needed immediate attention. It also emphasizes the significance of clear communication with stakeholders to ensure alignment and manage expectations.

Lesson learnt: Critical Path Analysis (CPA)

Critical Path Analysis (CPA) was crucial to my experience with construction projects in the oil and gas and mining sectors. By identifying the critical path, which represents the sequence of activities with the longest duration and no room for delay, I could focus on the most important tasks and allocate resources accordingly. Critical Path Analysis (CPA) is particularly useful in complex projects with multiple interdependent tasks.

Lessons learnt include identifying and focusing on the critical path, as any delay in these tasks will directly impact the project's overall timeline. CPA helps prioritize activities, allocate resources, and identify potential risks and bottlenecks. It also emphasizes the importance of regularly reviewing and updating the critical path analysis to account for any changes or adjustments in the project scope or schedule. CPA helped me understand the dependencies between activities and allowed for effective time management, ensuring that the project stayed on track and met its deadlines.

Overall, these tools and techniques, such as Responsibility Assignment Matrix (RAM), network diagrams, Gantt Charts, and Critical Path Analysis (CPA), have proven to be fundamental in successfully managing oil and gas and mining construction projects. They have taught me the importance of clear communication, efficient planning, and proactive risk management, leading to improved project outcomes.

Lesson Learned: Mind Mapping

Mind mapping is a creative and effective method for breaking down projects into smaller, manageable tasks.

During a recent project, I utilized mind mapping to brainstorm ideas and visually organize the project's various components. By starting with a central idea or goal in the center of the mind map, I was able to branch out and add subtasks, dependencies, and potential solutions. This technique allowed me to see the bigger picture while also focusing on the details. This method also assists you in getting a result very quickly.

One key lesson I learned from mind mapping was the importance of flexibility. As the project progressed and new information emerged, I was able to easily add or modify branches on the mind map, adapting the project plan accordingly. This flexibility helped me stay agile and respond to changing circumstances without feeling overwhelmed.

Another valuable lesson from mind mapping was the ability to identify potential bottlenecks or dependencies. I could easily spot areas where delays or issues might arise by visually representing the project components and their relationships. This allowed me to proactively address these challenges and allocate resources accordingly, ensuring a smooth project execution.

Lesson Learned: Affinity and Fishbone Diagrams

The affinity and fishbone diagrams are valuable tools for breaking down projects by identifying root causes and grouping related tasks.

Affinity diagrams help to categorize ideas and tasks into logical groups. This technique helped me identify common themes and patterns, making prioritising tasks and allocating resources easier. Through analysis, I identified key areas of focus and allocated resources accordingly.

Another technique I used was the fishbone diagram, also known as the Ishikawa diagram. This diagram helped me identify potential root causes for problems or delays in the project. I could pinpoint the underlying issues and take appropriate corrective actions by categorizing the causes into different branches, such as people, processes, or technology.

One important lesson I learned from using affinity and fishbone diagrams was the importance of involving the project team in the process. By facilitating collaborative brainstorming sessions and encouraging team members to contribute their ideas, we generated a comprehensive list of tasks and potential causes. This not only improved the accuracy of the diagrams but also fostered a sense of ownership and engagement among the team members.

Lesson Learned: Decision trees

I used decision trees with some projects to evaluate different options and their potential consequences. By mapping out different decision points and their associated outcomes, I could assess each option's risks and benefits.

This helped me make informed decisions and choose the project's most suitable course of action.

One key lesson I learned from using decision trees was the importance of considering various scenarios and their probabilities. By assigning probabilities to different outcomes, I could quantify the risks and make more objective decisions. This approach allowed me to prioritize tasks and allocate resources based on the likelihood and impact of different outcomes.

Another valuable lesson from using decision trees was the need for continuous monitoring and adjustment. As the project progressed, new information became available, and the probabilities of different outcomes changed. By regularly updating and revisiting the decision tree, I could adapt the project plan and make adjustments to ensure project success.

NOTE: If you use a standard WBS and compare the Organizational chart to a Resource chart, you could keep the Organizational chart levels down to department heads only.

For example, the scenario above makes sense if your Project Management team consists of 650 plus people or a large PMO. This allows each department to manage its own discipline resources.

Using mind mapping, affinity and fishbone diagrams, and decision trees greatly enhanced my ability to break down projects into manageable tasks, identify root causes, and make informed decisions. These techniques provided valuable insights and helped me plan and execute projects effectively, leading to successful outcomes.

116

CHAPTER 6

Implementing WBS & Other Structures

Assigning responsibilities and roles

Assigning responsibilities and roles is an essential step in project management. It ensures that every team member knows their part in the project and helps avoid confusion and overlapping tasks.

To assign roles and responsibilities, the project manager should start by creating a Work Breakdown Structure (WBS). A WBS is a hierarchical chart that breaks down the project into smaller, more manageable tasks. It helps identify all the activities required to complete the project and the resources needed.

Once the WBS is created, the project manager can assign each task to a specific team member. The team member responsible for a particular task should have the necessary skills, knowledge, and experience to complete it successfully. The project manager should also ensure that each team member clearly understands their role and responsibilities.

Ensuring that each team member knows the project objectives and timelines is essential. The project manager should provide regular updates on the project's progress and any changes that may affect the timeline or scope of work.

The project manager should also ensure each team member has the necessary resources to complete their tasks. This includes access to technology, equipment, and other tools required to complete the project successfully.

Clear communication is crucial in assigning roles and responsibilities successfully. The project manager should ensure that each team member knows who they report to and their point of contact for any questions or concerns.

Assigning roles and responsibilities is a critical step in project management. It ensures that every team member knows their part in the project and helps avoid confusion and overlapping tasks. A comprehensive WBS is essential for assigning tasks, and the project manager should ensure that each team member has the necessary resources to complete their tasks successfully. Clear communication is key to ensuring everyone knows the project objectives and timelines.

Monitoring and controlling the WBS

Monitoring and controlling the WBS is a crucial aspect of project management. It ensures that the project progresses according to plan and that potential issues are promptly identified and addressed. In this subchapter, we will explore the key elements of monitoring and controlling the WBS.

Establishing a baseline is the first step in monitoring and controlling the WBS. This baseline should include the project scope, schedule, budget, and quality objectives. The baseline serves as a reference point for measuring the project's progress and performance. Any deviation from the baseline should trigger an investigation into the root cause and corrective action.

Each team member must know their responsibilities, ensuring that these are meticulously measured and aligned with the project's primary objectives. When a project is delayed or has unanticipated difficulties, the lack of a baseline complicates everything and makes the situation highly unpleasant to all project participants. It becomes very unpleasant for the project manager.

NOTE: Creating a baseline is the most important step in the planning process before monitoring and controlling starts. There are quite a few project managers that neglect this critical step.

When the project is completed successfully, then there are no worries. There would be no consequences for not creating a baseline (most of the time).

However, if the project fails, things start going very wrong for the project team. Questions must be answered about why and how things went wrong. Not having a baseline does not assist the team at all.

The baseline itself should encompass a holistic view of the project, including its scope, schedule, budget, and quality objectives. It functions as a vital reference point, facilitating the measurement of the project's ongoing progress and performance in relation to the initial original plan. Hopefully, a meticulous change control regime was followed to ensure that "all" changes are part of the current baseline plan. I am mentioning this here, as I have seen various projects where some stakeholders request changes, but when things go wrong, they are nowhere to be seen when the proverbial starts hitting the fan.

We discuss change control in more depth later in this book. This means any deviations from the established project baseline warrant a thorough investigation to unearth the root cause (could use a fishbone diagram), followed by the implementation of corrective actions by the project team to realign the project with its original trajectory.

In summary, establishing a baseline is an essential phase in the project management process because it ensures that there is a clear and measurable standard against which the project's progress and performance can be evaluated. This proactive strategy not only aids in the detection of potential issues before they escalate, but also improves the project management process's overall accountability and openness.

The second step is to measure the project's performance against the baseline using metrics. These metrics should be aligned with the project objectives and provide insights into the project's health. Examples of metrics include schedule variance, cost variance, quality metrics, and risk metrics. The project team should track these metrics regularly and report them to the stakeholders.

The third step is to identify and manage risks. Risks are events or conditions that could impact the project's success. The project team should identify and assess risks early in the project lifecycle and develop mitigation strategies to minimize their impact. Risks should be monitored regularly, and mitigation strategies should be adjusted as necessary.

The fourth step is to monitor the project's scope. Scope creep is a common problem in projects, and it occurs when the project's scope expands beyond the original plan. The project team should monitor the scope regularly and ensure that the stakeholders approve any changes.

The fifth step is to manage project changes. Change management is the process of identifying, evaluating, and approving changes to the project's scope, schedule, budget, or quality objectives. Changes should be assessed for their impact on the project's baseline, and any necessary adjustments should be made.

In conclusion, monitoring and controlling the WBS is critical to project management. It enables the project team to promptly identify potential issues and take corrective action. By establishing a baseline, measuring performance using metrics, managing risks, monitoring scope, and managing changes, the project team can ensure the project's success.

Updating Changes the WBS

As a project team member, you may encounter situations where the project scope changes or new tasks are added to the project plan. In such cases, updating the Work Breakdown Structure (WBS) to reflect the changes accurately is essential. The WBS is a critical tool in project management that provides a visual representation of the project scope and helps to break down the project into manageable tasks. Here are some tips for updating the WBS effectively.

- **Review the project scope:** Before making any changes to the WBS, review the project scope to ensure that the changes align with the project goals. Identify the changes that need to be made and determine how they will affect the project timeline, budget, and resources.

- **Identify new tasks:** If new tasks need to be added to the project plan, identify them and determine where they fit into the WBS. Ensuring that the new tasks are relevant to the project scope and align with the project goals is essential.

- **Evaluate the impact of changes:** When making changes to the WBS, it is important to evaluate the impact on the project timeline, budget, and resources. Identify any potential risks and determine how they can be mitigated.

- **Update the WBS:** Once you have evaluated the impact of changes and identified new tasks, update the WBS to reflect the changes accurately. Ensure that the WBS is updated in a timely manner to avoid confusion and miscommunication.

- **Communicate changes:** Communicate the changes to the project team, stakeholders, and sponsors to ensure that everyone is on the same page. Provide an updated version of the WBS to all stakeholders and explain the changes that have been made.

In conclusion, updating the WBS is an essential part of project management. It helps to ensure that the project stays on track and aligns with the project scope and goals. As a project team member, it is important to review the project scope, identify new tasks, evaluate the impact of changes, update the WBS, and communicate changes to stakeholders. By following these steps, you can ensure that the WBS remains an effective tool in project management.

Communicating the WBS to stakeholders

Communicating the WBS to stakeholders is an essential aspect of project management. It is crucial to ensure that everyone involved in the project understands its scope, timelines, and deliverables. Here, I will provide insights into effectively communicating the WBS to stakeholders.

The first step in communicating the WBS to stakeholders is to **identify who the stakeholders are.** This includes the project team, sponsors, clients, and other relevant parties. Once the stakeholders are identified, creating a communication plan that outlines how and when the WBS will be communicated to them is crucial.

One effective way of communicating the WBS to stakeholders is **through visual aids such as diagrams and charts**. These visual aids provide a clear and concise overview of the project scope, timelines, and deliverables. It also helps stakeholders to understand the interdependencies between different tasks and how they fit into the overall project.

Another effective way of communicating the WBS to stakeholders is **through presentations**. Presentations provide an opportunity to discuss the WBS in detail and to answer any questions that stakeholders may have. This is particularly useful for complex projects where stakeholders may need more information to fully understand the project scope.

It is also important to ensure that the language used to communicate **the WBS is clear and concise**. Avoid using technical terms and jargon that may be unfamiliar to some stakeholders. Instead, use simple language that is easy to understand.

In conclusion, communicating the WBS to stakeholders is critical to project management. It helps to ensure that everyone involved in the project understands its scope, timelines, and deliverables. By identifying stakeholders, creating a communication plan, using visual aids, and clear language, project teams can effectively communicate the WBS to stakeholders.

Integrating the WBS with other project management tools

Integrating the WBS with other project management tools is essential for a successful project. The WBS provides a framework for the project, breaking it down into manageable pieces. However, it is not the only tool that project teams can use. Other project management tools can be used in conjunction with the WBS to ensure that the project is delivered on time, within budget, and to the required quality.

Project teams can use **the project schedule** as one of the most important tools. The project schedule is a timeline that outlines the tasks that need to be completed and when they need to be completed. The project schedule can be created using various project planning tools (Primavera, Open Plan, MS Project, Smartsheet's, Monday.com, etc.), including Gantt charts, critical path analysis, and network diagrams. The WBS can be used to inform the project schedule, with each work package being assigned a duration and a start and end date.

Risk management is another tool that can be used in conjunction with the WBS. Risk management involves identifying potential risks to the project and developing strategies to mitigate those risks. The WBS can be used to identify potential risks, with each work package being assessed for its likelihood of success and potential impact on the project if it fails. The risk management plan can then be developed to address these risks.

Project teams can also **use quality management tools** to ensure that the project meets the required quality standards. Quality management involves planning, executing, and controlling quality throughout the project. The WBS can be used to identify quality requirements for each work package, with quality control measures being put in place to ensure that these requirements are met.

Finally, project teams can **use communication tools** to ensure that stakeholders are kept informed of the project's progress. Communication tools can include project status reports, progress updates, and stakeholder meetings. The WBS can be used to inform these communication tools, with progress updates being provided for each work package.

There are various software tools pertaining to risks, quality, and planning, and I discuss the details of that in my risk management, scheduling, and quality management books.

In conclusion, integrating the WBS with other project management tools is essential for a successful project. The WBS provides a framework for the project, but it is not the only tool that project teams can use. By using other project management tools in conjunction with the WBS, project teams can ensure that the project is delivered on time, within budget, and to the required quality.

Key Tips: Implementing Structures

When implementing a Work Breakdown Structure (WBS), it is crucial to involve all stakeholders and team members in the process. Their input and expertise will ensure that the WBS accurately reflects the project's scope and objectives. By assigning responsibilities and accountability to specific roles will ensure that resources own their deliverables. By allocating these deliverables to key performance indicators (KPI) or key result areas (KRA) linked to the project will provide an effective measure.

Start by identifying the major deliverables and work packages that need to be completed for the project. Breaking down the project into smaller, manageable tasks will make it easier to assign responsibilities and track progress.

Use a hierarchical structure when developing the WBS to clearly define the relationship between different tasks and subtasks. This will help in understanding the dependencies and sequence of work.

Ensure that each work package in the WBS is measurable and has clear criteria for completion. This will help in tracking progress and evaluating the project's overall performance.

Consider using other structures, to assist you in managing your project. Using too many structures will keep everyone busy, which might not be necessary, and detract everyone from getting the project deliverables done. All these structures will help in assigning resources, clarifying roles and responsibilities, and ensuring effective communication within the project team. Use them wisely! I have used all of these structures before, but only I only used various of these structures for specific reasons during the project phases and life-cycles of a project. I will endeavour to include these structures when I write about specific processes and disciplines and delve into more detail around these structures and their outcomes.

125

Regularly review and update the WBS as the project progresses. As new information becomes available or changes occur, it is important to adjust the WBS accordingly to maintain accuracy and alignment with the project's objectives.

Communicate the WBS and other structures clearly to all team members and stakeholders. Make sure everyone understands their roles, responsibilities, and the overall project structure. This will promote collaboration, prevent misunderstandings, and enhance project efficiency.

Use project management software or tools to create and manage the WBS and other structures. These tools can simplify the process, improve collaboration, and provide real-time visibility into the project's progress.

Regularly monitor and track the project's progress against the WBS and other structures. This will help in identifying any deviations, addressing issues promptly, and ensuring that the project stays on track.

Finally, continuously learn and improve from each project's implementation of the WBS and other structures. Reflect on lessons learned, gather feedback from team members, and make adjustments for future projects to ensure ongoing success.

Lessons Learnt: Implementing Structures

Lesson 1: Define the Scope

Clearly define the project scope before creating the Work Breakdown Structure (WBS).

In one of my past projects that was under my guidance, we were tasked with developing a new training facility for a major client. The project team develop a original baseline WBS, schedule and budget for this project. Due to changes initiated (I must say implemented by bullying or overruling tactics) by various stakeholders, they had other expectations without fully understanding the project's scope. This resulted in a WBS that lacked clarity and was prone to changes without approval.

Lesson 2: Developing the WBS

During a project, we had a situation where the complete project team was not involved in creating the WBS due to commitment and time constraints. As a result, the stakeholders and leadership had limited understanding of the project's scope and how their individual work packages contributed to the overall project objectives. This lack of buy-in and understanding from the stakeholders and leadership caused miscommunication, delays, and poor project performance. I realized the importance of involving the project team in creating the WBS to foster ownership, improve communication, and ensure alignment with project goals.

NOTE: Always get your whole team involved with developing the WBS or at least review it (with it corresponding time and budget) before the project starts.

Lesson 3: Change Control

There was an initial well-defined WBS that accurately reflected the project's scope. However, as the project progressed, the client requested changes in campaign objectives and deliverables. Unfortunately, we neglected to get a reviewed and updated WBS approved by our stakeholders, leading to confusion, missed tasks, and a deviation from the original project plan. This experience re-iterated the importance of regularly reviewing and updating the WBS to accommodate changes in project requirements, ensuring alignment with the evolving project goals. All changes were managed and brought to the leadership and stakeholders to sign off, but to no avail.

The original baseline project charter was the most important document that was not managed with this project. It also demonstrates (I hope) that the leadership and stakeholders can also derail a project and throw the whole project team under the bus to CTA (cover their a….).

Lesson 4: Other Structures

This brings me to a few critical structures that we did put in place, Stakeholder and Communication breakdown structures (lots of internal and external stakeholders), but never got the approval of signoff due to responsibility and accountability changes to the leadership and stakeholders. One of the stakeholders was a consultant who did excellent work but was driving his own agenda and aligned the other stakeholders to his agenda without consulting the project team about the consequences.

Lesson 5: Time and Budget Control

As is quite evident, this projects time and budget was burnt quite early in the project, but I must say, it was a drop in the ocean, 0.0015% of the entire business plan budget. Funny enough, the amount we originally requested for this project is precisely where the budget ended up. So, all the line items (work) that the stakeholders did not approve during the feasibility phase, were requested during the execution of the project with obvious consequences. As they say, "Some projects are small but deadly."

An important lesson is to conduct a thorough scoping exercise with the stakeholders and the relevant leadership at the start of the next phase before creating and finalising the WBS, Schedule and Budget to ensure a comprehensive and accurate breakdown of project deliverables.

NOTE: Recheck your budget and time handed to you before you start the next phase of the project.

This is a step most project teams do not take. You will find that in some organizations, the teams that produce the design phase of a project are not the ones implementing the project.

As I mentioned, some stakeholders do not spend enough effort during these phases and sign off on the scope without proper due diligence.

Check the scope, budget, and schedule handed to you.

CHAPTER 7

Best Practices Using WBS & Other Structures

Defining clear project goals and objectives

Defining clear project goals and objectives is a crucial step towards the successful completion of any project. It allows the project team to focus on what is important and directs their efforts towards achieving specific outcomes. In this chapter, we will explore the importance of defining clear project goals and objectives and discuss some of the best practices for doing so.

Project goals and objectives are the foundation of any project. They provide a clear understanding of what the project aims to achieve and serve as a roadmap for the project team. Without clear goals and objectives, the project can easily become derailed, leading to delays, budget overruns, and even project failure.

To define clear project goals and objectives, involving all stakeholders in the process is important. This includes the project sponsor, project manager, project team, and any other relevant parties. By involving everyone in the process, you can ensure that the goals and objectives are aligned with the organization's overall strategy and that they are realistic and achievable.

When defining project goals and objectives, it is important to use the SMART criteria. SMART stands for Specific, Measurable, Achievable, Relevant, and Time-bound. This framework ensures that the goals and objectives are well-defined, quantifiable, realistic, and relevant and have a clear timeline for completion.

In addition to using the SMART criteria, it is important to ensure that the project goals and objectives are aligned with the project scope and deliverables. This means that they should be consistent with the project's requirements and constraints and should be achievable within the project's timeframe and budget.

Finally, it is essential to document the project goals and objectives and communicate them clearly to all stakeholders. This includes creating a project charter or a project plan that outlines the goals and objectives and the project scope, deliverables, timelines, and budget.

In conclusion, defining clear project goals and objectives is a critical step towards project success. By involving all stakeholders in the process, using the SMART criteria, aligning the goals and objectives with the project scope and deliverables, and documenting and communicating them clearly, the project team can ensure everyone is on the same page and working towards a common goal.

Here are some samples of SMART criteria one could use:

Discipline: Cost Control

Specific: Reduce project overhead costs by 10%.

Measurable: Track monthly expenditures against the baseline budget using cost management software.

Achievable: Implement cost-saving measures such as renegotiating contracts and optimizing resource allocation.

Relevant: Aligns with the organization's objective to improve profit margins on projects.

Time-bound: Achieve the cost reduction within the next fiscal quarter.

Discipline: Schedule Management

Specific: Decrease project delivery time by two weeks for the upcoming infrastructure project.

Measurable: Use Gantt charts and critical path analysis to monitor progress and identify any schedule slippage.

Achievable: Streamline processes and increase team shifts to ensure continuous work progress.

Relevant: Supports the company's reputation for timely delivery, enhancing customer satisfaction and competitive edge.

Time-bound: Complete the acceleration within the project's current lifecycle, before the end of the third quarter.

Discipline: Quality Assurance

Specific: Improve the quality rating of our construction deliverables by 15% as per the industry standard benchmarks.

Measurable: Conduct bi-weekly quality audits and gather feedback using standardized checklists.

Achievable: Provide additional training to the team and update quality control protocols.

Relevant: Ensures project outcomes meet the high standards expected by stakeholders and reduces the risk of rework.

Time-bound: Attain this improvement by the end of the current project phase.

Discipline: Risk Management

Specific: Identify and mitigate 90% of the high-priority risks in the project risk register.

Measurable: Review and update the risk register weekly, tracking mitigation actions and their effectiveness.

Achievable: Assign risk owners and develop robust mitigation plans for each identified high-priority risk.

Relevant: Directly contributes to the project's stability and predictability, which are critical for stakeholder confidence.

Time-bound: Complete risk mitigation actions within the first half of the project timeline.

Discipline: Stakeholder Engagement

Specific: Increase stakeholder satisfaction by 20% by the end of the project.

Measurable: Use surveys and feedback tools post each major milestone to measure satisfaction levels.

Achievable: Enhance communication plans and involve stakeholders in decision-making processes where appropriate.

Relevant: High stakeholder satisfaction can lead to repeat business and positive referrals, impacting the company's growth.

Time-bound: Reach the targeted satisfaction level before the project closing phase.

The project team can transform these and other SMART criteria into KPIs and KRAs for the project or organization.

Collaborating with team members

Collaborating with team members is one of the most critical components of any project. Through collaboration, team members can share ideas, insights, and knowledge, leading to better project outcomes. Let's delve into the importance of collaboration, how to collaborate effectively with team members, and the benefits that come with it.

The Importance of Collaboration

Collaboration is vital in project management because it allows team members to pool their resources, skills, and knowledge to achieve a common goal. Collaboration helps reduce individual team members' workload, leading to better productivity, efficiency, and effectiveness. Furthermore, collaboration enhances communication, which is essential in project management. It promotes transparency, accountability, and trust among team members, leading to better project outcomes.

How to Collaborate Effectively with Team Members

To collaborate effectively with team members, creating an environment that fosters collaboration is important. This involves establishing clear communication channels, setting expectations, and building trust among team members. Some of the ways to collaborate effectively with team members include:

> **1. Establishing clear roles and responsibilities:** Every team member should clearly understand their role and what is expected of them. This helps avoid confusion and ensures everyone is working towards the same goal.

> **2. Encouraging open communication:** Team members should be encouraged to share their ideas, insights, and concerns openly. This helps to promote a culture of transparency and accountability.

3. Building trust: Trust is essential in any collaborative effort. Team members should be encouraged to trust one another and to work towards building strong relationships.

Benefits of Collaboration

Collaboration has numerous benefits in project management. Some of the benefits include:

1. Improved project outcomes: Collaboration leads to the sharing of ideas, resources, and knowledge, leading to better project outcomes.

2. Increased productivity: Collaboration reduces the workload of individual team members, leading to increased productivity.

3. Enhanced communication: Collaboration promotes transparency, accountability, and trust among team members, leading to better communication.

Conclusion

Collaboration is vital in project management. It promotes transparency, accountability, and trust among team members, leading to better project outcomes. By establishing clear roles and responsibilities, encouraging open communication, and building trust among team members, project team members can collaborate effectively, leading to better productivity, efficiency, and effectiveness.

Adapting to changes in the project scope

Adapting to changes in the project scope is a crucial aspect of project management that can either make or break the success of a project. Projects are dynamic and can change at any time, making it necessary for project team members to be agile in their approach to project management. When changes occur in the project scope, it is important to approach them with a positive mindset and a willingness to adapt to new situations. In this subchapter, we will explore the importance of adapting to changes in the project scope and provide some tips on how to do so effectively.

Firstly, it is important to understand that changes in the project scope are a natural part of the project lifecycle. As such, project team members should be prepared to handle changes as they occur. When changes are made to the project scope, it is essential to assess the impact of the changes on the project schedule, budget and overall project goals. This assessment will help the team to determine whether the changes are feasible and what adjustments may be necessary.

Effective communication is also critical when adapting to changes in the project scope. Project team members should communicate the changes to all stakeholders and update project plans accordingly. It is important to keep all stakeholders informed of any changes and to seek their input and feedback where necessary.

Another critical aspect of adapting to changes is to remain flexible and open-minded. Project team members should be willing to make adjustments to project plans and work collaboratively to find solutions to any challenges that arise. This flexibility will ensure that the project stays on track and that the project goals are achieved.

In conclusion, adapting to changes in the project scope is a critical aspect of project management that requires agility, effective communication, and flexibility. By adopting a positive mindset and a willingness to adapt to new situations, project team members can ensure the project succeeds, even when changes occur.

Reviewing and assessing project progress

Reviewing and assessing project progress is an essential aspect of project management. This process involves evaluating the project's current status, identifying any potential issues, and determining whether the project is on track to meet its objectives. It is crucial for project team members to regularly review and assess project progress to ensure that the project is successful.

One effective way to assess project progress is to use a Work Breakdown Structure (WBS). The WBS is a hierarchical decomposition of the project into smaller, more manageable components. It provides a clear picture of the project's scope and helps the team members track progress against the project plan. By regularly reviewing the WBS, project team members can identify any areas that are falling behind schedule or require additional resources.

The Gantt chart is another structure that can be used to assess project progress. The Gantt chart visually represents the project schedule, which shows the project timeline, task dependencies, and resource allocation. By reviewing the Gantt chart regularly, project team members can see if the project is progressing as planned or if there are any delays that need to be addressed.

In addition to using these structures, project team members should also conduct regular status meetings to review project progress. During these meetings, team members should discuss any issues or risks that have arisen and determine the appropriate actions to address them. The team should also review the project plan and ensure that it is still relevant and achievable.

Finally, project team members should gather feedback from stakeholders and assess their satisfaction with the project's progress. This feedback can be used to identify areas for improvement and ensure that the project is meeting its goals.

In conclusion, reviewing and assessing project progress is critical to the success of any project. By using structures like the WBS and Gantt chart, conducting regular status meetings, and gathering feedback from stakeholders, project team members can ensure that the project is on track to meet its objectives.

Continuous improvement of project management practices

Project management is a continuously evolving field that requires a proactive approach to stay relevant and effective. As project team members, it is important to recognize that there is always room for improvement in project management practices. With the increasing complexity of projects, it is crucial to have a structured approach to ensure that the project is completed on time, within budget, and with the desired quality.

To achieve continuous improvement, it is essential to assess the current practices and identify areas that need improvement. One effective tool to use is the Work Breakdown Structure (WBS), which helps in breaking down a project into smaller, more manageable tasks. The WBS allows project team members to identify dependencies, allocate resources, and monitor progress effectively.

Apart from the WBS, other structures such as the Critical Path Method (CPM) and Gantt charts can be used to improve project management practices. The CPM helps identify the critical path of a project, which is the sequence of tasks that must be completed in a specific order to ensure the project is completed on time. Gantt charts provide a visual representation of the project schedule and help in monitoring progress.

Continuous improvement in project management practices also requires a review of lessons learned from previous projects. By understanding what worked well and what did not, project team members can implement best practices and avoid repeating past mistakes. It is also important to stay up-to-date with the latest project management tools and methodologies, such as Agile and Scrum, to ensure that projects are completed efficiently and effectively.

In conclusion, continuous improvement of project management practices is critical to the success of any project. By using tools such as the WBS, CPM, and Gantt charts, and by reviewing lessons learned, project team members can identify areas for improvement and implement best practices. By staying up-to-date with the latest project management methodologies, project team members can ensure that projects are completed on time, within budget, and with the desired quality.

Key Tips: Best Practices WBS

As a project manager, it is essential to understand the importance of a well-defined Work Breakdown Structure (WBS) and other project structures. These structures help organize and manage project tasks efficiently.

When creating a WBS, start by identifying the project's major deliverables. Break down these deliverables into smaller, more manageable work packages. This hierarchical structure will provide clarity and a systematic approach to project execution.

Ensure that each work package in the WBS has a clear and specific scope, well-defined objectives, and measurable outcomes. This will help in assigning responsibilities and tracking progress effectively.

While creating the WBS, involve the project team members and stakeholders. Their input and perspective can add valuable insights and ensure that all relevant tasks and activities are included in the project scope.

Consider using other project management structures in conjunction with the WBS. These structures help define work packages, roles, and responsibilities, highlight dependencies, and enhance communication and coordination among team members.

Regularly review and update the WBS and other project structures as the project progresses. This will help accommodate any changes, identify potential risks, and ensure that the project remains on track. By reviewing the WBS that is aligned to the scope, Scope Management and Tracking (Scope Creep) can be reduced and managed. Allocation of new WBS elements can create a major deviation in the project deliverables.

It is crucial to maintain consistency and standardization in the WBS and other project structures across different projects within an organization. This will enable easier integration of resources, knowledge sharing, and better project portfolio management.

NOTE: When using a WBS in portfolio management, one could show the overall company, then the various sites, and then type of projects. This could be split further by quarter, year, or planning period.

A Schedule or Gantt chart would demonstrate this clearly (per WBS structure). A project life cycle planning period in portfolio management can run over 20 to 50 years.

Use appropriate software tools and technologies to create and manage the WBS and other project structures. These tools can help automate processes, facilitate collaboration, and generate reports for better decision-making.

Communicate the WBS and other project structures effectively to the project team members, stakeholders, and other relevant parties. This will ensure a shared understanding of project goals, expectations, and deliverables.

Finally, continuously learn and improve your skills in creating and using WBS and other project structures. Stay updated with industry best practices, attend training programs, and seek feedback from experienced project managers. This will help enhance your project management capabilities and achieve successful project outcomes.

Lessons Learnt: Best Practices WBS

Some of these items seem repetitive, but they are listed to reinforce the concept of setting up a proper WBS. Most of the time, project managers list some activities, allocate headings to those activities, and use that as their WBS. This is not the way to develop a WBS, as some team members do not read the scope properly, and various visualizations of the WBS exist. I repeat some lessons, just in case you have not read certain sections of this book.

Lesson 1: Clearly Define the WBS

I have learned that it is crucial to clearly define the work breakdown structure (WBS) in project management. Breaking down the project into smaller, manageable components makes it easier to assign roles and responsibilities, track progress, and identify potential risks. For example, in a software development project, we used a WBS to divide the project into phases such as requirements gathering, design, development, testing, and deployment. This helped us allocate resources effectively and ensure that all necessary tasks were accounted for. It also assisted us with the timing of the various team members and the required interaction and interfaces. This becomes very crucial during the testing and deployment phase.

Lesson 2: Involve Stakeholders in the WBS Development

In my experience, involving stakeholders in the development of the work breakdown structure (WBS) leads to better project outcomes. By including input from key stakeholders, such as clients, end-users, and subject matter experts, we were able to ensure that all project requirements were considered and addressed. For instance, we engaged architects, engineers, and the client in a construction project to develop the WBS together. This enabled us to align the project scope with the client's expectations and avoid costly rework.

Lesson 3: Use Hierarchical Structures to Organize Tasks

I have learned that using hierarchical structures, such as the WBS, helps organize tasks and facilitate efficient project management. Breaking down the project into smaller components and organizing them in a logical order allows for better resource allocation and task scheduling. As an example, in an engineering company strategy and updated marketing campaign project, we used a hierarchical structure to categorize tasks into broad categories like strategy development, listed all activities required over a 90-day rollout of the "new" company strategy, created new content, design, employee engagement, and promotion. This helped us prioritize activities and allocate resources accordingly, ensuring a timely strategic project delivery.

Lesson 4: Leverage Other Structures in Addition to the WBS

While the WBS is a powerful tool, I have realized the importance of leveraging other structures in project management. For example, using organizational breakdown structures (OBS) helps define roles and responsibilities within the project team. By clearly identifying the reporting lines, the relationships and communication channels, we were able to streamline decision-making processes and avoid conflicts. Additionally, using a resource breakdown structure (RBS) allowed us to allocate resources effectively, ensuring that the right people were assigned to the right tasks based on their skills and availability.

Lesson 5: Review and Update the WBS and Other Structures

Lastly, I have learned the importance of regularly reviewing and updating the work breakdown structure (WBS) and other structures throughout the project lifecycle. As the project progresses and new information becomes available, it is crucial to adapt and refine these structures to reflect the evolving needs and scope. By conducting regular reviews, we can identify any required gaps or (Scope) changes, ensuring that the project remains on track and aligned with the overall objectives. For example, in a construction project, we regularly reviewed and updated the WBS to accommodate changes in customer requirements and trends, allowing us to deliver a successful product.

NOTE: Be aware that some clients do not know what they want. "OR" They do not read the original scope document provided to them, even though they signed it off. Stakeholder engagement is crucial during the design phases to ensure that the end product is what they expect to be delivered.

Here is an example:

Let's say you want to build a house from scratch, you design the house with a designer and a drafty (or Architect), but you keep changing your mind about what you want. Plus, by adding all the nice-to-have's, you'll find that the house is becoming too expensive. Setting up the WBS or Estimate structure correctly, will assist with change control. I used this as an example in the estimating (EBS) section.

NOTE: The ratios between the various structures you use in your project can be: 1 = One, M = Many.

1 : 1,

M : M,

M : 1, and/or 1 : M.

CHAPTER 8

WBS Structures Conclusion

Recap of key takeaways

As a project team member, you play an important role in the success of a project. To ensure that your project runs smoothly, it is essential to have a clear understanding of the project's scope, goals, and objectives. In PART I, we have discussed the importance of the work breakdown structure (WBS) and other structures in breaking down projects into manageable tasks.

Here are some of the key takeaways:

1. Work Breakdown Structure (**WBS**) is a **hierarchical decomposition** of the project into smaller, more manageable components.

2. **WBS** helps to **identify all the tasks** that need to be completed and ensures that each task is assigned to the right team member.

3. WBS also helps to **identify the critical path of the project**, which is the sequence of tasks that must be completed on time to ensure project success. This would be according to me the Schedule Breakdown Structure (SBS).

4. **Other structures**, (we will only name a few) such as Organizational Breakdown Structure (OBS), Estimate Breakdown Structure (EBS), Cost Breakdown Structure (CBS), and Resource Breakdown Structure (RBS), can also be used to **break down projects into manageable** components.

5. **OBS** helps to identify the **roles and responsibilities** of different team members and ensures that each team member knows what is expected of them.

145

6. **EBS** helps to **split the costs down to finite detail** that is required in a bottom-up type estimate.

7. **CBS** guides the team of where the **monies are allocated and controlled**. Contingency, management reserve, and escalation is always a contentious item in the WBS. Use your organizations guides on cost control as that will assist you in developing the right control budgets for your projects.

8. **RBS** helps to **identify the resources needed** for each task and ensures that those resources are available when needed.

9. **Communication is key to project success**. Regular communication between team members and stakeholders helps to ensure that everyone is on the same page and that any issues or concerns are addressed promptly.

10. Finally, it is important to **monitor and control** the project to ensure that the **project stays on track**. Regular monitoring helps to identify any issues or risks that may arise, and control measures can be put in place to mitigate those risks.

In conclusion, breaking down projects into manageable tasks using WBS and other structures is essential for project success. As a project team member, it is important to have a clear understanding of these structures and to communicate effectively with other team members and stakeholders. By following the key takeaways outlined here, you can ensure that your project runs smoothly and achieves its objectives.

Final thoughts and recommendations.

Congratulations! You have reached the end of PART I of this comprehensive guide to WBS and other project structures. By now, you should have a good understanding of how to break down projects into manageable pieces, assign responsibilities, and track progress. As a project team member, you play a critical role in ensuring that your project is completed on time, within budget, and to the satisfaction of your stakeholders.

Before we conclude, here are some final thoughts and recommendations to help you make the most of what you have learned:

1. Always start with the end in mind. Before you begin breaking down your project, make sure you have a clear understanding of what you are trying to achieve. This will help you ensure that your project is aligned with your organization's goals and objectives.

NOTE: Start with the handover (all projects have a handover phase) and work your way to the beginning or the start of a project. Once you have this basic thread, you will be fine.

---

2. Use the right tool for the job. You can use many different project structures, from WBS to Gantt charts to Agile sprints. Make sure you choose the one that is best suited to your project and your team's needs.

3. Communicate, communicate, communicate. Effective communication is essential to the success of any project. Make sure you keep your stakeholders informed of your progress, any issues you encounter, and any changes you need to make.

4. Embrace change. Projects are rarely straightforward, and unexpected challenges are bound to arise. Embrace the opportunity to be flexible and adapt to changing circumstances.

5. Learn from your mistakes. No project is perfect, and there will always be something you could have done better. Take the time to reflect on what worked well and what didn't, and use that knowledge to improve your processes in the future.

NOTE: Stick to the KISS principle.

Keep it Simple Simon or is it 'Stupid'. I should know, as my dad always used 'stupid', never Simon. That helped me to start using common sense when making decisions.

Do not over-complicate your life; projects are complicated and stressful enough. The more complicated you make your project during the setup to impress your peers and leadership, the more issues and possible hassles you will have during your project life cycle.

Deliver what is needed for the project and no more. Nobody has the time (or money) to manage additional issues.

To conclude Part I, the essence of breaking down projects into manageable pieces is essential to the success of any project. By using the right tools and techniques, communicating effectively, and embracing change, you can ensure that your projects are completed on time, within budget, and to the satisfaction of your stakeholders. We hope that this guide has been helpful to you and wish you all the best in your future project management endeavours.

PART II – WBS DICTIONARY

CHAPTER 9

Introduction to the WBS Dictionary

Evolution and History of the WBS Dictionary

The importance of using a Work Breakdown Structure Dictionary (WBSD) cannot be overstated when it comes to effective project management. The WBS dictionary serves as a vital tool that helps project teams navigate the complexities of their projects by providing a comprehensive breakdown of project deliverables, activities, and key information. This subchapter, "Evolution and History of the WBS Dictionary," aims to shed light on the origins and development of this indispensable resource.

The concept of the WBS dictionary can be traced back to the early 1950s when the United States Department of Defense initiated the Program Evaluation Review Technique (PERT). PERT aimed to streamline the management of complex projects, particularly in the defense and aerospace sectors. As part of this effort, the WBS dictionary was introduced to describe project tasks and their associated attributes systematically.

Over the years, the WBS dictionary has evolved and adapted to the changing landscape of project management. Initially, it was a simple document that included only basic information about project activities. However, as projects became more complex and diverse, the need for a more comprehensive and standardized approach to project documentation emerged. This led to the development of software tools and templates that enabled project teams to create and maintain WBS dictionaries more efficiently.

Today, the WBS dictionary has become an integral part of project management methodologies such as the Project Management Institute's (PMI) Project Management Body of Knowledge (PMBOK). It is recognized as a valuable resource for clearly defining project scope, facilitating communication, and ensuring consistency across all project phases.

Using a WBS dictionary offers numerous benefits to project teams. Firstly, it provides a common language that promotes effective communication among team members, stakeholders, and clients. By clearly defining project deliverables, activities, and dependencies, the WBS dictionary helps prevent misunderstandings and ensures that everyone is on the same page.

Secondly, the WBS dictionary enhances project planning and control. It allows project managers to allocate resources, estimate costs, and establish realistic timelines accurately. Moreover, it enables effective monitoring and tracking of project progress, as deviations from the original plan can be easily identified and addressed.

Thirdly, the WBSD must be developed during a project's design stages / phases. As mentioned earlier, don't over-complicate your life, and don't try to impress your boss. Some projects are complicated enough. Only develop a breakdown structure if you require the structure to manage specifics in a project environment.

NOTE: Remember, don't forget the KISS principle.

Lastly, the WBS dictionary serves as a valuable knowledge repository for future projects. Organizations can leverage their previous experiences to improve project outcomes and reduce risks by documenting lessons learned, best practices, and historical data.

In conclusion, the evolution and history of the WBS dictionary demonstrates its significance as a project management tool. By providing a comprehensive breakdown of project activities and associated information, it facilitates effective communication, enhances planning and control, and fosters knowledge sharing. As project teams strive for success in an ever-evolving business landscape, utilizing a WBS dictionary becomes an essential practice for achieving project objectives efficiently and effectively.

Overview of the WBS Dictionary

In today's fast-paced business environment, where projects are becoming increasingly complex, having a clear and well-defined WBS is crucial. A WBS is a powerful tool that helps project teams break down their project scope into manageable and actionable components. It provides a systematic approach to organizing and understanding project deliverables, activities, and milestones. Simply creating a WBS is not enough - it is equally important to use a WBS dictionary.

However, creating and maintaining a comprehensive WBS can be a daunting task, especially when dealing with large-scale projects or multiple teams. This is where a WBS dictionary comes into play. A WBS dictionary is a companion document that provides detailed descriptions of the elements within the WBS, including their purpose, scope, and responsibilities.

Let's discuss the process of creating and utilizing a WBS dictionary. We will explore the various components of a WBS and explain how to leverage the dictionary to enhance project planning, communication, and execution.

Whether you are a project manager, team member, or stakeholder, PART II will be your go-to resource for effectively understanding and implementing

WBS dictionaries.

Throughout the chapters, you will learn how to create clear and concise descriptions for each WBS element, ensuring that everyone on the project team has a shared understanding of the project scope and objectives. This will facilitate communication and alignment, reducing the risk of misunderstandings and potential project delays.

Furthermore, PART II will provide practical tips and best practices for maintaining and updating the WBS dictionary throughout the project lifecycle. We will address common challenges that project teams face when managing a WBS dictionary and offer strategies for overcoming them.

By the end of PART II, you will have a solid understanding of the importance of using a WBS dictionary and how it can streamline project management processes. You will be equipped with the knowledge and tools necessary to create and maintain an effective WBS dictionary for your projects, improving overall project success rates and team collaboration.

So, let's unravel the intricacies of Work Breakdown Structures and empower project teams to reach new heights of project management excellence.

CHAPTER 10

Understanding Work Breakdown Structure Dictionaries

Definition and Purpose of a WBS Dictionary

As mentioned, a WBS alone is insufficient to ensure effective project management. This is where a WBS Dictionary (WBSD) comes into play to assist in keeping all the various structures aligned in one dictionary.

The WBS Dictionary can be defined as a key companion document to the WBS, containing detailed information about each component or work package within the structure. It serves as a comprehensive reference guide for the project team, providing vital information about the scope, requirements, and certain information about other breakdown structures and expectations associated with each element.

The purpose of a WBS Dictionary is twofold: communication and clarity. It serves as a means to communicate the project's scope and requirements to all stakeholders, ensuring a shared understanding of the project's objectives. By providing a detailed description of each work package, including its purpose, deliverables, and associated resources, the dictionary enables effective communication and collaboration among team members and stakeholders.

Furthermore, the WBS Dictionary enhances clarity by reducing ambiguity and minimizing misunderstandings. It eliminates any confusion by documenting the specific details, assumptions, and constraints related to each work package.

This promotes a common understanding of project scope, expectations, and dependencies, thereby reducing the risk of errors, rework, and scope creep.

NOTE: Be conscious of having too much information in your WBS dictionary. Some data is required, and other information can put you on a path of having an information overload. Only put in what is needed.

Remember, whatever you add to your WBSD, you will track in another document that requires updating and control. Try not to duplicate everything.

Developing an adequately detailed WBSD for your project will reflect and address the complete scope of a project in much more detail than a standard WBS.

The importance of using a WBS Dictionary cannot be overstated. It acts as a central repository of project information, facilitating effective decision-making, problem-solving, and risk management. It enables project teams to track progress, identify dependencies, allocate resources, and manage changes efficiently. Moreover, the dictionary serves as a valuable reference for future projects, allowing organizations to leverage their previous experiences, best practices, and lessons learned.

Components of a WBS Dictionary

One of the key components of a WBS dictionary is the identification of project deliverables. Deliverables are the tangible outcomes or results that the project aims to accomplish. The WBS dictionary lists each deliverable and provides a detailed description, including its purpose, functionality, and any specific requirements. This ensures that the project team has a clear understanding of what needs to be achieved and helps prevent scope creep.

Another important component of a WBS dictionary is the decomposition of deliverables into work packages. Work packages are smaller, manageable units of work that can be assigned to specific team members or departments. The dictionary provides a breakdown of each deliverable into its corresponding work packages, along with their descriptions, durations, and dependencies. This allows for better resource allocation, task assignment, and tracking of progress.

The WBS dictionary also includes information about project activities and tasks. It provides a detailed description of each task, including its start and end dates, dependencies, and responsible parties. This helps project teams to identify potential bottlenecks, allocate resources effectively, and ensure that tasks are completed on time and within budget.

NOTE: The WBS dictionary must not duplicate your project schedule activities. The WBS and the WBS dictionary are presented at a higher level than the planned project activities and deliverables or Schedule Brakdown Structure (SBS).

Here is a WBS dictionary card that you could use to get the basic information together to develop a WBS dictionary. I will supply additional information that you could use to develop a WBS dictionary in PART III of this book.

Importance of Using a WBS Dictionary

A WBS dictionary is a companion document to the WBS that provides detailed information about each component or task. It serves as a reference guide for the project team, ensuring everyone is on the same page regarding the work to be done.

Here are some key reasons why using a WBS dictionary is essential:

1. **Clarity and Consistency:** A WBS dictionary eliminates ambiguity by clearly defining each task, including its purpose, deliverables, and any dependencies. This ensures that all team members have a consistent understanding of the work breakdown structure, reducing the chances of misunderstandings or miscommunication.

2. **Resource Allocation:** Project managers can accurately allocate resources to each task with a WBS dictionary. It provides information on the skills, materials, and equipment required for each component, enabling efficient resource planning and allocation. This prevents overloading or underutilization of resources, leading to better project execution.

3. **Change Management:** Projects are dynamic, and changes are inevitable. A WBS dictionary facilitates effective change management by capturing the impact of any modifications to the project scope or tasks. It helps in evaluating the ripple effect of changes on other components, allowing project teams to make informed decisions and adjust timelines or resources accordingly.

4. **Stakeholder Communication:** A WBS dictionary is a valuable tool for communicating with project stakeholders. It provides a common language and understanding of the project's work breakdown structure, making it easier to discuss progress, issues, and expectations. Stakeholders can refer to the dictionary to gain insights into specific tasks, fostering better collaboration and engagement.

5. **Knowledge Transfer:** A WBS dictionary serves as a repository of project knowledge. It captures the lessons learned, best practices, and historical data associated with each task. This knowledge can be

invaluable for future projects, allowing teams to leverage past experiences, avoid pitfalls, and improve overall project performance.

A WBS dictionary is a vital component of successful project management, and project teams can enhance their efficiency, minimize risks, and achieve project objectives with greater ease.

WBS Number	WBS Name	WBS Author
WBS Predecessor	WBS Description:	
WBS LOE:		
Must Start:		
Must Finish:	Assumptions and Constraints	
Date Completed:		

Figure 20 - Basic Components of a WBS Dictionary

Benefits of Using a WBS Dictionary

A WBS dictionary serves as a reference guide for the project team. Here are some key benefits of using a WBS dictionary:

1. **Improved Communication:** A WBS dictionary enhances communication among project team members. By providing clear descriptions and definitions of each work package, it eliminates any ambiguity or confusion. This leads to effective communication and a shared understanding of project requirements, resulting in better collaboration and teamwork.

2. **Enhanced Project Planning:** The detailed information in a WBS dictionary allows project teams to plan and schedule tasks more accurately. It provides insights into the resources, time, and effort required for each work package, enabling better resource allocation and project scheduling. This leads to improved project planning and increased chances of meeting project deadlines.

3. **Better Risk Management:** A WBS dictionary helps identify potential risks associated with each work package. With a clear understanding of the project scope, project teams can identify and address risks more effectively. By having detailed information about dependencies, constraints, and deliverables, project teams can proactively manage risks, reducing the likelihood of negative impacts on project outcomes.

4. **Facilitates Change Management:** Projects often face changes, whether it's scope changes, resource changes, or schedule changes. A WBS dictionary makes it easier to manage these changes. With a clear understanding of the project scope and dependencies, project teams can assess the impact of changes more accurately and make informed decisions. This ensures that changes are implemented smoothly and without disrupting project progress.

Project teams can use a WBS dictionary to ensure a common understanding of project requirements, leading to successful project execution and delivery.

Linking the WBS and WBS Dictionary

One of the primary benefits of using a WBS Dictionary is clarity. Linking the WBS and WBS Dictionary eliminates any ambiguity by providing clear definitions, descriptions, and specifications for each work package. This clarity helps prevent misunderstandings and ensures that all team members have a shared understanding of their responsibilities. It also facilitates effective communication with stakeholders, enabling project managers to explain the project's scope and deliverables concisely and accurately.

Another key advantage of linking the WBS and WBS Dictionary is traceability. The WBS Dictionary allows project teams to trace each work package back to its respective higher-level deliverables. This traceability helps in identifying dependencies and understanding the impact of any changes or delays. By having a clear understanding of how each element fits into the larger structure, project teams can proactively manage risks and make informed decisions.

Furthermore, the WBS Dictionary provides a valuable resource for future projects. As project teams complete their work, they accumulate knowledge and lessons learned. By documenting this information in the WBS Dictionary, organizations can benefit from the insights gained and avoid repeating mistakes in future projects. This knowledge transfer ensures continuous improvement and increases the efficiency and effectiveness of project management processes.

Key Tips: Understanding a WBSD

The Work Breakdown Structure (WBS) is a fundamental tool that helps project teams organize and manage their work effectively. However, the true power of the WBS lies in its detailed documentation and the use of a Work Breakdown Structure dictionary.

1. **Definition and Purpose of a WBS Dictionary:** A Work Breakdown Structure dictionary is a comprehensive document that accompanies the WBS. It contains detailed descriptions, definitions, and attributes for each work package or element within the WBS. The dictionary aims to ensure consistent understanding and communication among project team members, stakeholders, and other relevant parties.

2. **Enhancing Clarity and Communication:** One of the primary benefits of using a Work Breakdown Structure dictionary is that it enhances clarity and communication within the project team. Team members can easily understand and discuss project tasks by providing a common vocabulary and clear definitions for each work package. This reduces misunderstandings and facilitates effective collaboration.

3. **Enabling Consistent Estimation and Planning:** Another key advantage of utilizing a Work Breakdown Structure dictionary is that it enables consistent estimation and planning. With detailed descriptions of work packages, project teams can accurately estimate the effort, duration, and resources required for each task. This leads to more realistic project schedules and budgets.

4. **Supporting Project Control and Reporting:** A Work Breakdown Structure dictionary also plays a vital role in project control and reporting. Project managers can track progress, identify bottlenecks, and generate accurate status reports by documenting specific attributes such as responsible parties, dependencies, and milestones. This helps ensure

that projects stay on track and meet their objectives.

5. **Facilitating Knowledge Transfer and Lessons Learned:** Lastly, a well-maintained Work Breakdown Structure dictionary serves as a valuable knowledge repository for future projects. It captures lessons learned, best practices, and historical data, which can be easily accessed for future reference. This promotes continuous improvement and streamlines future project planning and execution.

By utilizing the WBSD tool effectively, teams can enhance clarity, communication, estimation, planning, control, reporting, and knowledge transfer. Investing time and effort in creating and maintaining a comprehensive WBS dictionary is a wise decision that will ultimately lead to project success and improved project management practices.

Lessons Learnt: Understanding a WBSD

In my years in project management, I've learned that breaking down a project using a Work Breakdown Structure (WBS) is crucial for clarity and manageability. But the real game-changer has been the integration of a Work Breakdown Structure Dictionary (WBSD). This tool has transformed how my team and I approach project tasks, providing a detailed reference that clarifies scope, deliverables, and dependencies.

Here are the lessons I've learned from using a WBSD:

Enhanced Communication: The WBSD has been instrumental in aligning the team's understanding of tasks, leading to improved coordination and the resolution of conflicts.

Consistency and Standardization: It has standardized our approach by defining terms and guidelines, streamlining processes, and improving the accuracy of our project plans. See the chapter about developing a WBSD.

Accurate Estimations: With clear guidelines, estimating resources, timelines, and costs has become more reliable, resulting in realistic project targets.

Knowledge Repository: The WBSD is an invaluable historical record for onboarding new team members and leveraging past insights for continuous improvement.

In essence, the WBSD is not just a document; it's the backbone of project clarity and efficiency. It ensures that every team member, present and future, can contribute effectively, making it an indispensable part of our project management toolkit.

CHAPTER 11

The WBS Dictionary

Create a WBS Dictionary

The WBS dictionary is a document that provides detailed information about each component of the project's WBS. It serves as a reference guide for project team members and stakeholders to understand each work package's scope, deliverables, and requirements.

The WBS dictionary typically includes (as a minimum) the following information for each work package:

- **Work Package Identifier:** Each work package is assigned a unique identifier, such as a number or code, to differentiate it from other components.

- **Work Package Description:** A brief description of the work package, including its purpose, objectives, and deliverables. This description should clearly define the scope of work and provide enough information for team members to understand what needs to be accomplished.

- **Work Package Scope:** This section outlines the boundaries and limitations of the work package. It defines what is included and excluded from the work package's scope, ensuring no ambiguity regarding the responsibilities and deliverables.

- **Work Package Dependencies:** Any dependencies or relationships the work package has with other project components should be documented here. This helps in understanding the sequence and interdependencies between work packages.

- **Milestones and Deliverables:** Key milestones and deliverables associated with the work package should be clearly identified. These milestones serve as checkpoints to track the progress of the project and ensure that each deliverable is completed on time and meets the required quality standards.

- **Responsible Party:** The individual or team responsible for executing the work package should be specified. Assigning clear responsibilities ensures accountability and helps track progress and resolve any issues or concerns that may arise.

- **Resource Requirements:** The resources, such as human resources, equipment, materials or software, needed to complete the work package should be identified. This includes the quantity, availability, and any specific skills or qualifications required for the resources.

- **Work Package Duration:** The estimated time required to complete the work package should be documented. This helps in scheduling and resource allocation, ensuring that the project stays on track and can be completed within the defined timeframe.

Identifying Key Deliverables and Work Packages

Creating a Work Breakdown Structure (WBS) dictionary is essential to project management. It helps in identifying key deliverables and work packages, which are crucial components of the project.

Key Deliverables: Key deliverables refer to the final output or results that are expected to be delivered at the end of the project. These deliverables represent the tangible outcomes or products that the project aims to achieve. Identifying key deliverables is important as it provides a clear understanding of what needs to be accomplished and helps set project objectives.

To create a WBS dictionary, key deliverables are listed as the top-level or higher elements. These deliverables are typically broken down into smaller, more manageable components known as work packages.

Work Packages: Work packages are the smallest units of work in a project. They represent the specific activities or tasks that need to be completed to achieve the key deliverables. Work packages are further broken down into smaller, more detailed tasks, enabling better planning, resource allocation, and tracking of progress.

When creating the WBS dictionary, each work package is assigned a unique identifier, description, and associated resources. The identifier helps in referencing and tracking the work package throughout the project lifecycle. The description provides a clear understanding of the work involved in the package, including its scope, objectives, and expected outcomes. Additionally, the resources associated with each work package include the individuals or teams responsible for its execution and any required materials, equipment, or tools.

The WBS dictionary serves as a comprehensive reference document that provides detailed information about each key deliverable and work package. It helps project managers, team members, and stakeholders understand each component's scope, requirements, and dependencies. This information is crucial for effective project planning, scheduling, budgeting, and resource allocation.

Defining Work Package Descriptions

The purpose of defining work package descriptions in the WBS dictionary is to ensure a common understanding among project team members about the specific tasks and activities that need to be completed to achieve project goals. Project managers can effectively assign responsibilities, allocate resources, and track progress by clearly defining the work packages.

The work package descriptions in the WBS dictionary should be concise, yet comprehensive enough to provide a clear understanding of the work involved. Each description should include the following key elements:

- **Scope:** Clearly define the boundaries and objectives of the work package. This helps in understanding what is included and what is excluded from the scope of the package.

- **Deliverables:** Specify the tangible outputs or outcomes that are expected to be produced as a result of completing the work package. This helps in setting expectations and measuring progress.

- **Activities:** Break down the work package into specific tasks and activities that need to be completed. This helps in understanding the sequence and dependencies of the work.

- **Resources:** Identify the resources required to complete the work package, including personnel, equipment, materials, and facilities. This helps in allocating resources effectively and avoiding any resource constraints.

- **Timeline:** Define the start and end dates for the work package, along with any key milestones or deadlines. This helps in managing the project schedule and tracking progress.

Creating detailed work package descriptions in the WBS dictionary is crucial for effective project management. It ensures that everyone involved in the project clearly understands the work to be performed, the expected outcomes, and the resources needed. This clarity helps avoid misunderstandings, reduce risks, and improve overall project performance.

Assigning Responsibility Codes and Resources

Responsibility codes are unique identifiers assigned to individuals or teams responsible for completing specific tasks or work packages. These codes help track and assign accountability for each project component.

Assigning responsibility codes involves creating a hierarchical structure that reflects the organizational structure of the project team. This allows for easy identification and tracking of responsibilities at different levels, from project managers to individual team members.

Resources, however, are the people, equipment, materials, or software needed to complete the project. Assigning resources to work packages involves identifying the resources required for each task or deliverable and ensuring availability.

By assigning responsibility codes and resources, project managers can effectively allocate tasks to the appropriate individuals or teams and ensure that the necessary resources are available to complete the work. This improves coordination, communication, and overall project efficiency, ultimately leading to successful project outcomes.

Establishing Dependencies and Predecessors

One of the crucial aspects of developing a WBS dictionary is establishing dependencies and predecessors. Dependencies refer to the relationships between different project activities or tasks, where the completion of one task is dependent on the completion of another task. Identifying and understanding these dependencies is crucial for effective project planning and scheduling. By establishing dependencies, project managers can determine which tasks need to be completed before others can start, ensuring a logical sequence of activities.

Predecessors, on the other hand, are the tasks that come before a particular task in the project schedule. They are the activities that directly influence or impact the start or finish of a subsequent task. Predecessors are essential for determining the order in which tasks need to be executed and the overall project timeline.

To establish dependencies and predecessors in the WBS dictionary, project managers should consider the following steps:

- **Identify all the tasks and activities involved in the project:** Start by listing all the tasks and activities required to complete the project. Break down larger deliverables into smaller, manageable components.

- **Determine the dependencies between tasks:** Analyze each task and identify its relationship with other tasks. Determine if a task depends on another task's completion or if it can be executed concurrently.

- **Define the type of dependencies:** Once dependencies are identified, categorize them based on their type. There are four common types of dependencies: finish-to-start (FS), start-to-start (SS), finish-to-finish (FF), and start-to-finish (SF).

- **Document the dependencies and predecessors:** In the WBS dictionary, create a section specifically for dependencies and predecessors. Document each task, its dependencies, and the corresponding predecessors. Include relevant information such as task names, task numbers, and descriptions.

- **Update and review the dependencies regularly:** As the project progresses, dependencies may change. It is crucial to continuously update and review the dependencies to ensure accuracy and alignment with the project's evolving needs.

Project managers can effectively plan and schedule project activities by establishing dependencies and predecessors in the WBS dictionary. This enables them to allocate resources efficiently, manage risks, and ensure the successful completion of the project within the defined timeline.

Updating and Maintaining the WBSD

Updating and maintaining the WBS dictionary is a crucial aspect of project management. The WBS dictionary serves as a comprehensive document that provides detailed information about each work package within the Work Breakdown Structure (WBS). It helps in creating a common understanding and clear communication among project team members, stakeholders, and other project participants.

The process of updating the WBS dictionary involves capturing any changes or updates to the project scope, deliverables, or objectives. This can be done through regular review meetings, discussions, or through the use of project management software. It is important to involve all relevant stakeholders during the update process to capture all necessary information accurately.

The WBS dictionary should be updated whenever there are changes to the project, such as new work packages, modifications to existing work packages, or changes in project requirements. These updates may occur as a result of scope changes, feedback from stakeholders, or lessons learned from previous phases of the project. It is essential to keep the WBS dictionary up to date to ensure that all project participants have access to the most current and accurate information.

Maintaining the WBS dictionary involves ensuring that the document remains accurate, complete, and easily accessible throughout the project lifecycle. This includes updating changes, adding new information, and removing outdated or irrelevant details. It is important to have a designated person or team responsible for maintaining the WBS dictionary and overseeing the update process.

Regularly reviewing and updating the WBS dictionary helps manage project risks and identify potential scope creep. By maintaining an accurate and up-to-date WBS dictionary, project managers can effectively track progress, identify dependencies, and allocate resources efficiently. It also helps monitor project performance, identify gaps or issues, and make informed decisions during project execution.

NOTE: Create a simple work breakdown structure dictionary.

To manage and control a WBSD, focus on time, cost, and a person who will take responsibility for the quality of each WBSD element.

Key Tips: Creating a WBS Dictionary

A WBS is a powerful tool that helps teams plan, execute, and control projects effectively. However, it is not enough to simply create a WBS; developing a comprehensive WBS dictionary is equally important.

A WBS dictionary serves as a repository of all the key information related to the project's deliverables and their associated work packages. It provides a clear and concise description of each component, including its purpose, scope, and requirements. By having a centralized and detailed source of information, project teams can effectively communicate, collaborate, and ensure consistency throughout the project lifecycle.

One of the key benefits of using a WBS dictionary is that it enhances communication and understanding among project team members. It eliminates confusion and ambiguity by providing a common language and a shared understanding of the project's objectives. This facilitates effective communication and collaboration, especially when dealing with complex or large-scale projects involving multiple team members and stakeholders.

Another advantage of a WBS dictionary is that it improves project planning and estimation. By breaking down the project's scope into manageable work packages, teams can accurately estimate each component's time, resources, and costs. This allows for better project planning and ensures that all necessary resources are allocated appropriately.

When creating a WBS dictionary, there are several key tips to keep in mind. First, ensure that each work package is clearly defined and described. Use concise and unambiguous language to avoid misinterpretation or confusion. Additionally, include any relevant technical specifications, dependencies, and constraints to provide a comprehensive understanding of each component.

It is also crucial to regularly update the WBS dictionary throughout the project lifecycle. As the project evolves, new deliverables may be added, scope may change, or dependencies may emerge. By maintaining an up-to-date WBS dictionary, teams can adapt to these changes and ensure that everyone has access to the most accurate and current information.

NOTE: Do not overcomplicate developing a work breakdown structure dictionary.

Only focus on what is necessary for a successful project.

Make sure your stakeholders (clients, especially) are aware of this fact.

Lessons Learnt: Creating a WBS Dictionary

A WBS serves as a foundational tool that helps organize and manage project tasks efficiently. However, the true power of a WBS can only be harnessed when it is accompanied by a comprehensive dictionary.

A WBS dictionary is essentially a repository of all the terms, definitions, and descriptions related to the various elements of a WBS. It acts as a common language for the project team, enabling effective communication and understanding across all members. Without a dictionary, there is a risk of misinterpretation and confusion, leading to delays, errors, and overall project inefficiency.

One of the key lessons learned in creating a WBS dictionary is the need for collaboration among the project team. Each team member brings a unique perspective and expertise, and their input is invaluable in accurately defining and documenting the WBS elements. Regular meetings and brainstorming sessions can facilitate this collaborative effort, ensuring that all team members are on the same page.

Another important lesson is the significance of simplicity and clarity in the dictionary entries. The purpose of a WBS dictionary is to provide a concise and easily understandable description of each element. Complicated jargon or ambiguous terms should be avoided, as they can lead to further confusion. It is crucial to strike a balance between providing sufficient information and keeping the entries concise and straightforward.

Additionally, maintaining consistency in terminology is key to the effectiveness of a WBS dictionary. The project team should establish clear guidelines and standards for defining and documenting the WBS elements. Consistency ensures that everyone is using the same terminology, reducing the chances of misunderstanding and facilitating smoother communication.

Furthermore, it is essential to keep the WBS dictionary updated throughout the project lifecycle. As the project progresses and new tasks or elements are added, the dictionary should be revised and expanded accordingly. Regular reviews and revisions ensure that the dictionary remains relevant and up-to-date, supporting the project team in their day-to-day activities.

NOTE: I will keep harping on this note.

I have seen projects (not mine) go horribly wrong by not remembering or using the KISS principle.

There is no need to impress your boss unnecessarily.

CHAPTER 12

Using a Work Breakdown Structure Dictionary

Navigating the WBS Dictionary

The WBS dictionary is a key tool in project management that provides detailed information about each element in the WBS. It serves as a reference guide for project stakeholders to understand the various components, deliverables, and activities within the project.

When navigating the WBS dictionary, it is crucial to understand its structure and content clearly. The following points highlight the main aspects of navigating the WBS dictionary:

- **Organizational Hierarchy:** The WBS dictionary follows the same hierarchical structure as the WBS itself. It starts with the highest level of the project and breaks down into lower levels, providing more detailed information at each level. Navigating the WBS dictionary involves moving through the levels to locate specific elements and associated information.

- **Element Identification:** Each element in the WBS dictionary is uniquely identified, usually by a code or number, which corresponds to the corresponding element in the WBS. This identification system allows for easy navigation and cross-referencing between the WBS and its dictionary.

- **Element Description:** The WBS dictionary provides a detailed description of each element. This description includes the scope, objectives, deliverables, and activities associated with that particular element. Navigating the WBS dictionary involves reading and understanding the element's descriptions to understand the project's components correctly and comprehensively.

- **Associated Information:** Apart from the description, the WBS dictionary may include additional information for each element. This information can encompass resource requirements, cost estimates, schedule constraints, quality standards, and any other relevant details. Navigating the WBS dictionary involves exploring these associated pieces of information to gain a deeper understanding of the project's requirements.

- **Cross-Referencing:** The WBS dictionary serves as a cross-reference tool for the project. It allows stakeholders to trace the dependencies and relationships between different elements. By navigating the WBS dictionary, one can easily identify the predecessors and successors of a particular element, enabling a better understanding of the project's sequence and interdependencies.

- **Updates and Revisions:** Throughout the project lifecycle, the WBS dictionary may undergo updates and revisions as the project progresses. Navigating the WBS dictionary involves keeping track of these changes to ensure that the information remains up-to-date and accurate. Regularly reviewing the WBS dictionary helps in maintaining project alignment and ensuring that all stakeholders are aware of any modifications made.

Navigating the WBS dictionary is essential for project management. It helps stakeholders understand the project's components, deliverables, and activities in detail. By following the organizational hierarchy, identifying elements, exploring descriptions and associated information, cross-referencing, and staying updated, project teams can effectively utilize the WBS dictionary as a valuable resource throughout the project lifecycle.

Interpreting Work Package Descriptions

Interpreting Work Package Descriptions is crucial to creating a Work Breakdown Structure (WBS) dictionary. The WBS dictionary serves as a comprehensive reference guide that provides detailed information about each work package included in the WBS.

- **Understand the purpose and scope of each work package.** A work package is a deliverable-oriented component of the project that defines the smallest unit of work within the WBS. It clearly describes the work to be performed, the expected outcomes, and any specific requirements or constraints associated with it.

- **Analyze & identify the elements that define the work package.** This involves breaking down the description into smaller, manageable tasks and understanding the dependencies and relationships between them.

- **Provide a concise title that represents the work package.** It should also provide a detailed description of the work to be performed, including the objectives, deliverables, and any associated milestones or deadlines.

- **Outline the required resources.** Resources such as personnel, equipment, or materials, needed to complete the work package successfully. This information is crucial for resource allocation and planning purposes.

- **Understanding any dependencies or relationships:** Relationships between different work packages help project managers identify the sequence in which work packages need to be executed and ensure that any dependencies are properly managed.

- **Include specific constraints or requirements:** Constraints or requirements must be considered during the execution of the work package. This could include factors such as budget limitations, regulatory compliance, or technical specifications.

Overall, interpreting Work Package Descriptions is essential for creating a comprehensive WBS dictionary. By analyzing and understanding the information provided in each work package description, project managers and team members can effectively plan, execute, and monitor the project's progress.

Understanding Responsibility Codes and Resources

Creating a WBS dictionary is an essential step in project management as it provides a detailed description of the work breakdown structure (WBS) components. It serves as a reference guide for the project team, stakeholders, and other parties involved in the project. A WBS dictionary enhances communication, ensures clarity, and helps in assigning responsibilities and resources effectively.

Understanding Responsibility Codes:

- Responsibility codes are identifiers assigned to specific tasks or work packages within the WBS.
- These codes help identify who is responsible for completing each task or work package.
- Responsibility codes are often represented by unique alphanumeric characters or symbols.
- They provide clarity and accountability by clearly defining the roles and responsibilities of individuals or teams involved in the project.
- Responsibility codes can be customized based on the project's needs, organizational structure, or industry standards.

Understanding Resources:

- Resources refer to the people, materials, equipment, and facilities required to complete the project.

- The WBS dictionary includes a detailed list of resources needed for each task or work package.

- It specifies the type, quantity, and availability of resources required.

- Resources can be categorized into human, physical, and financial resources.

- Human resources include project team members, contractors, and subject matter experts.

- Physical resources include tools, machinery, infrastructure, or any tangible items required for the project.

- Financial resources include budget allocations, funding sources, or any monetary requirements.

By including responsibility codes and resources in the WBS dictionary, project managers can ensure that the right people are assigned to the appropriate tasks. This helps in optimizing resource allocation, avoiding conflicts, and ensuring smooth workflow throughout the project. The WBS dictionary acts as a central repository of information, facilitating effective communication and coordination among the project team members.

Analyzing Dependencies and Predecessors

Understanding dependencies and predecessors is crucial for the successful execution of any project. These two concepts play a significant role in the development and implementation of a WBS dictionary:

- **Identifying dependencies:** One of the first steps in creating a comprehensive WBS dictionary is to analyze and identify the dependencies between different work packages or tasks. Dependencies can be categorized as either internal (within the project) or external (dependent on factors outside the project's control). This analysis helps determine the sequence in which tasks must be executed and how they rely on each other.

- **Defining dependencies:** Once the dependencies are identified, it is important to define them clearly within the WBS dictionary. This includes specifying the type of relationship between tasks, such as finish-to-start, start-to-start, finish-to-finish, or start-to-finish. By defining these dependencies, the project team can understand how one task may impact another and plan accordingly.

- **Documenting predecessors:** Predecessors are tasks that need to be completed before another task can start. It is crucial to document the predecessors for each task in the WBS dictionary. This information helps in scheduling and resource allocation, as it ensures that the necessary prerequisites are met before moving on to the next task.

- **Analyzing critical path:** The critical path is the longest sequence of dependent tasks that determine the project's overall duration. Analyzing the dependencies and predecessors in the WBS dictionary can help identify the critical path, enabling project managers to focus their efforts on these critical tasks to ensure timely project completion.

- **Mitigating risks:** By analyzing dependencies and predecessors, project teams can identify potential risks and plan mitigation strategies accordingly. For example, if a critical task has a high dependency on an external factor, the team can develop contingency plans or establish alternative methods to reduce the impact of any delays or disruptions.

- **Ensuring accuracy and clarity:** The WBS dictionary should provide accurate and clear information about dependencies and predecessors. It should be easily understood by all stakeholders, including project team members, contractors, and clients. Clear documentation helps in effective communication, decision-making, and coordination among project participants.

- **Regular updates:** Dependencies and predecessors may change throughout the project lifecycle due to various factors such as scope changes, resource constraints, or external influences. Therefore, it is important to regularly review and update the WBS dictionary to reflect any changes in dependencies. This ensures that the project plan remains up-to-date and aligned with the evolving project requirements.

- **Tools and techniques:** Various tools and techniques can be used to analyze dependencies and predecessors, such as network diagrams, Gantt charts, and project management software. These tools help visualize and manage the relationships between tasks, enabling project teams to effectively analyze and plan their project activities.

Leveraging the WBSD for Reporting and Communication

One of the key benefits of having a well-defined WBS dictionary is leveraging it for reporting and communication purposes.

Reporting plays a crucial role in keeping stakeholders informed about the project's progress, status, and any potential issues or risks. By utilizing the WBS dictionary, project managers can easily generate reports that provide a comprehensive overview of the project's scope, objectives, and milestones. The dictionary helps in categorizing and listing all the deliverables, activities, and tasks included in the WBS, enabling project managers to track progress accurately.

Moreover, the WBS dictionary facilitates effective communication among project team members and stakeholders. It serves as a common reference point that ensures everyone involved in the project shares a clear understanding of the project's scope and objectives. The dictionary provides a detailed description of each component of the WBS, including its purpose, requirements, and dependencies. This information aids in avoiding misunderstandings and misinterpretations, thus enhancing collaboration and reducing the likelihood of project delays or mistakes.

In addition to aiding communication within the project team, the WBS dictionary is also valuable for external communication with stakeholders. It enables project managers to effectively communicate project requirements, expectations, and progress to clients, sponsors, or other external parties. The dictionary provides a structured and organized format that all stakeholders can easily share and understand, fostering transparency and trust in the project.

Overall, leveraging the WBS dictionary for reporting and communication purposes is vital for successful project management. It ensures accurate reporting of project progress and facilitates effective communication among project team members and stakeholders. This comprehensive reference tool allows project managers to streamline project execution, mitigate risks, and maintain alignment with project objectives.

Key Tips: Using a WBS Dictionary

One of the most crucial tools at a project team's disposal is the Work Breakdown Structure (WBS) dictionary. This powerful resource serves as a comprehensive guide to understanding and organizing the various components of a project. Here, we explore the importance of using a WBS dictionary and provide key tips on how to effectively utilize this invaluable tool.

First and foremost, a WBS dictionary acts as a centralized repository for all project-related information. It provides a detailed breakdown of the project's scope, objectives, deliverables, and tasks, allowing team members to understand their roles and responsibilities clearly. By having this information readily available, project teams can ensure that everyone is on the same page, reducing the risk of miscommunication and increasing overall productivity.

One of the key benefits of using a WBS dictionary is its ability to facilitate effective project planning and scheduling. The team can create a realistic timeline, allocate resources efficiently, and identify potential bottlenecks or dependencies by breaking down the project into manageable tasks and sub-tasks. The dictionary also allows for easy tracking and monitoring of progress, enabling project managers to make informed decisions and take corrective actions when necessary.

Furthermore, a WBS dictionary serves as a valuable reference tool throughout the project's lifecycle. It provides the entire team a common language and terminology, ensuring clarity and consistency in all project-related communications. Additionally, it allows for seamless collaboration and knowledge transfer, as new team members can quickly get up to speed by referring to the dictionary. This saves time and enhances the team's overall efficiency and effectiveness.

To make the most out of a WBS dictionary, following a few key tips is essential. Firstly, ensure that the dictionary is regularly updated and maintained to reflect any changes or updates in the project. This will ensure that the information remains accurate and relevant. Secondly, encourage all team members to actively contribute to the dictionary by providing updates or suggestions. This collaborative effort will help in capturing a broader perspective and foster a sense of ownership among the team.

Lastly, promote the use of the WBS dictionary as a go-to resource for all project-related queries. By emphasizing its importance and accessibility, team members will be more inclined to refer to it and rely on it as a trusted source of information.

Lessons Learnt: Using a WBS Dictionary

Reflecting on my extensive experience in project management and the implementation of Work Breakdown Structure (WBS) dictionaries, I've learnt several lessons that have significantly enhanced my approach to managing complex projects.

A WBS dictionary serves as a companion to the WBS, providing detailed descriptions and definitions of each element or task within the structure. It acts as a reference guide for the project team, ensuring a common understanding of the project scope and minimizing ambiguity. Here, we explore the importance of using a Work Breakdown Structure dictionary and the valuable lessons learnt from its implementation.

One of the key lessons learnt in using a WBSD is the enhancement of communication and collaboration within the project team. By clearly defining and documenting each element in the WBS, team members can easily refer to the dictionary to understand their tasks' purpose, requirements, and dependencies. This shared knowledge fosters effective communication, reducing misunderstandings and promoting efficient collaboration.

Another lesson learnt is the facilitation of project planning and scheduling. A WBSD allows project managers to allocate resources and estimate timeframes accurately. By having detailed descriptions of each task, the project team can better estimate the effort required, allocate resources accordingly, and create a realistic project schedule. This helps in avoiding underestimation or overallocation of resources, leading to improved project planning and successful execution.

Furthermore, a WBSD aids in risk identification and management. A comprehensive understanding of each task and its dependencies can identify potential risks early on. This enables the project team to proactively address risks and develop mitigation strategies for project success. The WBSD acts as a valuable resource for risk assessment and management, providing insights into potential challenges and allowing for timely and effective risk response.

The dictionary provides a common language and understanding for the project team by documenting and defining each element within the WBS. The lessons learnt showcase the immense value it brings, enabling successful project execution and delivering desired project outcomes.

In summary, the lessons I've learned from implementing a WBS dictionary have enhanced my capabilities in project management and contributed significantly to the successful delivery of projects. The WBSD has proven to be an indispensable tool in my arsenal, enabling clear communication, efficient planning, effective risk management, and fostering a unified team approach.

CHAPTER 13

Best Practices for Implementing a WBSD

Engaging Stakeholders in the WBSD

Engaging stakeholders in the WBSD is essential to ensure their active participation, alignment, and understanding of the project scope and deliverables. Here are the best practices for engaging stakeholders:

- **Identify and involve key stakeholders:** Identify all relevant stakeholders with a vested interest in the project and involve them in creating and reviewing the WBSD. This includes project team members, sponsors, clients, end-users, subject matter experts, and other relevant parties.

- **Communicate the purpose and benefits:** Clearly communicate the purpose and benefits of the WBSD to stakeholders, highlighting how their engagement and input will contribute to project success. Emphasize that their involvement will ensure accurate project scoping, resource allocation, and accountability.

- **Conduct stakeholder interviews or workshops:** Organize interviews or workshops with stakeholders to gather their insights, perspectives, and expectations regarding the project scope and deliverables. These sessions can help foster collaboration, clarify requirements, and identify potential risks or dependencies.

- **Solicit feedback and validate information:** Regularly seek feedback from stakeholders on the content, accuracy, and usability of the WBSD. Validate the information provided by stakeholders by cross-referencing it with other project documentation and conducting thorough reviews.

187

- **Provide training and support:** Offer training sessions or workshops to stakeholders to familiarize them with the WBSD structure, terminology, and usage. Provide ongoing support and guidance to address any questions or concerns they may have during the implementation and maintenance of the WBSD.

- **Maintain an open feedback loop:** Encourage stakeholders to provide continuous feedback on the WBSD as the project progresses. Actively listen to their suggestions, concerns, and updates to ensure the WBSD remains relevant, up-to-date, and aligned with evolving project requirements.

- **Document stakeholder decisions and agreements:** Keep a record of stakeholder decisions and agreements related to the WBSD. This documentation helps maintain transparency, accountability, and serves as a point of reference during project execution and change management processes.

- **Regularly review and update the WBSD:** Schedule regular reviews and updates of the WBSD with stakeholders to incorporate any changes in project scope, requirements, or deliverables. This ensures that the WBSD remains accurate, reflects the project reality, and is aligned with stakeholder expectations.

By implementing these best practices, project managers can engage stakeholders effectively in the WBSD process, ensuring a shared understanding of project scope and deliverables. This collaboration promotes stakeholder satisfaction, minimizes misunderstandings, and increases the likelihood of project success.

Ensuring Consistency and Accuracy in WBSD

Ensuring consistency and accuracy in the WBSD is of utmost importance to maintain project integrity and facilitate smooth communication among team members. Let's discuss some best practices to achieve consistency and accuracy:

- **Standardize terminology:** Use consistent language and terminology throughout the WBSD to avoid confusion and misinterpretation. Establish guidelines for naming conventions, abbreviations, and acronyms to maintain clarity and ensure everyone understands the components and deliverables.

- **Clearly define deliverables:** Each deliverable in the WBSD should have a clear and concise description that leaves no room for ambiguity. Clearly define each deliverable's scope, objectives, and expected outcomes to prevent any misunderstandings or misinterpretations.

- **Assign unique identifiers:** Assigning unique identifiers, such as codes or numbers, to each component and deliverable in the WBSD helps in maintaining accuracy and consistency. These identifiers should be easily recognizable and searchable, allowing team members to locate specific items in the dictionary quickly.

- **Establish a change management process:** Implement a well-defined change management process to handle any modifications or updates to the WBSD. Clearly document the steps required to request, review, approve, and implement changes, ensuring that all relevant stakeholders are involved in the process. This helps in maintaining the accuracy and consistency of the WBSD as it evolves throughout the project lifecycle.

- **Conduct regular reviews and audits:** Periodically review and audit the WBSD to identify any inconsistencies, inaccuracies, or gaps. This can be done by involving key stakeholders, project managers, and subject matter experts. Regular reviews help keep the WBSD current and aligned with the evolving project requirements.

- **Document assumptions and dependencies:** Clearly document any assumptions or dependencies associated with the components and deliverables in the WBSD. This ensures that team members are aware of any specific requirements or constraints that may impact their work. Regularly review and update these assumptions and dependencies to maintain accuracy and consistency.

- **Provide training and guidance:** Conduct training sessions to familiarize team members with the WBSD and its proper usage. Ensure that everyone understands the importance of consistency and accuracy in the dictionary. Provide guidelines and support materials, such as templates and examples, to assist team members in correctly entering data into the WBSD.

- **Foster collaboration and communication:** Encourage collaboration among team members and promote open communication channels to address any concerns or questions related to the WBSD. Regularly communicate updates, changes, and clarifications regarding the WBSD to ensure everyone is working with the most accurate and consistent information.

By following these best practices, project managers can ensure that the WBSD remains a reliable and valuable resource throughout the project lifecycle, aiding in effective project planning, execution, and control.

Training Team Members on WBSD

Training team members on the Work Breakdown Structure Dictionary (WBSD) is crucial for successfully implementing and utilizing the WBSD. By providing proper training, team members can better understand the purpose and benefits of the WBSD and how to effectively use and update it. Here are some best practices to consider when training team members on the WBSD:

- **Clearly communicate the importance of the WBSD:** Explain to

190

team members why the WBSD is essential for organizing and managing project deliverables. Highlight how it can improve project planning, communication, and collaboration.

- **Provide an overview of the WBSD:** Begin the training by providing an introduction to the WBSD, including its structure, key elements, and how it relates to the overall project management process. This will help team members grasp the concept and purpose of the WBSD.

- **Demonstrate how to use the WBSD:** Show team members how to navigate and use the WBSD effectively. Provide step-by-step instructions on accessing and updating the WBSD, adding new deliverables, and modifying existing entries. Use examples and practical exercises to reinforce learning.

- **Emphasize the importance of accurate & consistent data entry:** Stress the significance of maintaining accurate and up-to-date information in the WBSD. Explain how inconsistent or incomplete data can impact project planning, resource allocation, and decision-making. Share guidelines for data entry and encourage team members to double-check their inputs.

- **Encourage collaboration and communication:** Highlight the collaborative nature of the WBSD and encourage team members to actively participate in its development and maintenance. Foster a culture of open communication and teamwork, where team members can share their insights, ask questions, and provide feedback on the WBSD.

- **Provide ongoing support and resources:** Offer continuous support to team members by providing training materials, user guides, and access to additional resources such as FAQs or a help desk. Establish a point of contact for any questions or issues related to the WBSD. Regularly assess the team's understanding and proficiency with the WBSD, and provide additional training sessions or refresher courses as needed.

- **Foster a learning environment:** Encourage team members to continuously improve their knowledge and skills related to the WBSD. Promote a learning culture by organizing workshops, webinars, or lunch-and-learn sessions where team members can share their experiences, best practices, and lessons learned.

By following these best practices, organizations can ensure that their team members are well-equipped to utilize the WBSD effectively, leading to improved project outcomes and overall project management success.

Integrating the WBSD with PM Software

Integrating the Work Breakdown Structure Dictionary (WBSD) with project management software can greatly enhance the efficiency and effectiveness of project planning and execution. By combining the comprehensive information captured in the WBSD with the capabilities of PM software, project managers can streamline processes, improve communication, and ensure project success. Here are some best practices for integrating the WBSD with PM software:

- **Select compatible software:** Choose project management software that allows for easy integration and customization of the WBSD. Look for software that supports importing and exporting capabilities and the ability to create custom fields and templates.
- **Align WBSD codes with software:** Ensure that the WBSD codes and categories are aligned with the software's structure and terminology. This will facilitate seamless integration and reduce confusion during data transfer.
- **Import the WBSD:** Import the WBSD into the PM software to establish a centralized repository of project information. This will

enable easy access to project-specific details, such as deliverables, activities, resources, and dependencies.

- **Link WBSD elements to tasks:** Link WBSD elements to specific tasks or activities within the PM software. This linkage will provide clear traceability between the project plan and the corresponding breakdown structure elements, allowing for better progress tracking and monitoring.

- **Utilize custom fields:** Take advantage of custom fields within the PM software to capture additional information from the WBSD. This could include attributes such as cost estimates, duration, responsible parties, or any other relevant details that the WBSD does not directly cover.

- **Regularly update the WBSD:** Keep the WBSD up to date by synchronizing it with the PM software. Any changes made in the software, such as new tasks or modifications to existing ones, should be reflected in the WBSD to maintain accuracy and consistency across the project.

- **Leverage reporting capabilities:** Leverage the reporting capabilities of the PM software to generate customized reports based on the WBSD data. This will help track progress, identify bottlenecks, and make informed decisions throughout the project lifecycle.

- **Train project team members:** Provide training to project team members on how to effectively use the integrated WBSD and PM software. This will ensure that everyone understands the importance of the WBSD and how to leverage its integration with the software for improved project management.

Integrating the WBSD with PM software is a powerful approach that can enhance project planning, execution, and control. By following these best practices, project managers can maximize the benefits of both the WBSD and the software, leading to more successful project outcomes.

Continuous Improvement and Lessons Learned

The ability to continuously improve is essential for success. As a project team, it is crucial to embrace the mindset of continuous improvement and learn from past experiences.

Continuous improvement and lessons learned play a crucial role in the successful implementation of a Work Breakdown Structure Dictionary (WBSD). As organizations strive to enhance their project management processes, it is essential to establish practices that facilitate continuous improvement and allow teams to learn from their experiences. This section will delve into the best practices for implementing a WBSD with a specific focus on continuous improvement and lessons learnt:

- **Regularly review and update the WBSD:** The WBSD should not be considered a static document. It should be regularly reviewed and updated to reflect the evolving needs and requirements of the projects. This ensures that the WBSD remains relevant and effective in guiding project managers and teams.

- **Encourage feedback and suggestions:** Creating an environment that encourages feedback and suggestions from project team members fosters a culture of continuous improvement. By actively seeking input, organizations can identify areas for improvement and implement necessary changes to the WBSD.

- **Conduct post-project reviews:** After the completion of each project, conducting post-project reviews is vital to capturing lessons learned. These reviews provide an opportunity to identify what worked well and what can be improved in future projects. The findings from these reviews can be used to update the WBSD, ensuring that lessons learned are incorporated into future project planning and execution.

- **Document and share lessons learned:** Lessons learned should be documented systematically and shared with relevant stakeholders. This could be done through a lessons learned repository or knowledge management system. By sharing lessons learned, organizations can avoid repeating mistakes, capitalize on successful strategies, and continuously improve their project management practices.

- **Establish a feedback loop:** To promote continuous improvement, it is crucial to establish a feedback loop between project managers and team members. This can be achieved through regular project status meetings, one-on-one discussions, or online collaboration platforms. By actively seeking feedback, project managers can identify areas for improvement in the WBSD and address any concerns or challenges faced by the team.

- **Provide training and support:** Continuous improvement requires a skilled and knowledgeable workforce. Organizations should invest in training programs to enhance project management skills and educate team members on the proper use of the WBSD. Additionally, providing ongoing support and guidance to project managers and teams can facilitate the implementation and continuous improvement of the WBSD.

By following these best practices, organizations can ensure that their WBSD remains a valuable tool for project management, enabling continuous improvement and fostering a culture of learning from past experiences.

Key Tips: Best Practices

Many project teams often overlook the importance of using a WBS dictionary as a companion tool to the WBS. A WBS dictionary provides a detailed explanation of the components and deliverables of the WBS, making it an indispensable asset for effective project planning and execution.

1. **Understand the Purpose of a WBS Dictionary:** Before implementing a WBS dictionary, it is important for project teams to understand its purpose. A WBS dictionary serves as a repository of information related to each WBS element, including its description, deliverables, responsible parties, and dependencies. It provides clarity and ensures consistent understanding among team members.

2. **Document WBS Elements Consistently:** Consistency is key when documenting WBS elements in the dictionary. Use a standardized format and ensure that all team members follow the same guidelines. This will enhance communication, reduce confusion, and facilitate easier tracking of project progress.

3. **Include Detailed Descriptions:** Each WBS element in the dictionary should have a detailed description that clearly explains its purpose, scope, and objectives. This will help team members understand the significance of each element and contribute effectively towards its completion.

4. **Define Deliverables and Milestones:** Clearly define the deliverables and milestones associated with each WBS element. This will enable project teams to track progress, identify dependencies, and allocate resources accordingly.

5. **Assign Responsibilities:** Assigning responsibilities to specific team members for each WBS element is essential for accountability and efficient project management. Clearly document the responsible parties within the WBS dictionary to avoid confusion and ensure smooth execution.

6. **Update and Review Regularly:** A WBS dictionary is a dynamic document that should be updated and reviewed regularly throughout the project lifecycle. As the project progresses, new deliverables may arise or existing ones may change. Regular updates and reviews will keep the WBS dictionary current and relevant.

Implementing a WBS dictionary is a best practice for project teams aiming to streamline project management processes. By following these key tips, project teams can ensure consistency, clarity, and effective communication within their project teams. A well-maintained WBS dictionary will be a valuable tool for achieving project success.

Lessons Learnt: Best Practices

For project teams seeking to enhance their project management practices, mastering the art of utilizing a WBS dictionary is crucial.

A WBS dictionary serves as a comprehensive reference guide that provides detailed information about each component of the WBS. It acts as a bridge between the project team and stakeholders, ensuring a common understanding of project scope, tasks, and deliverables. By utilizing a WBS dictionary, project teams can achieve the following benefits:

1. **Clarity and Consistency:** A well-defined WBS dictionary provides clear and concise descriptions of work packages, enabling project team members to have a shared understanding of the project objectives. It helps eliminate ambiguity and minimizes the chances of miscommunication or misunderstandings.

2. **Improved Communication:** The WBS dictionary acts as a common language for the project team and stakeholders. It facilitates effective communication by ensuring that everyone is on the same page regarding project scope, requirements, and expectations.

3. **Enhanced Planning and Resource Allocation:** The WBS dictionary provides critical information about the resources required for each work package. Project managers can make informed decisions regarding resource allocation, scheduling, and budgeting by having a clear overview.

To make the most of a WBS dictionary, project teams should consider the following best practices for implementation:

1. **Collaboration and Input:** Involve all relevant stakeholders, including project team members, subject matter experts, and clients, in the creation and maintenance of the WBS dictionary. Encourage open dialogue, feedback, and contributions to ensure accuracy and completeness.

2. **Standardization:** Establish a standardized format for the WBS dictionary. Use consistent terminology, definitions, and formatting across all projects. This will enhance clarity, ease of use, and understanding.

3. **Regular Updates:** The WBS dictionary should be a living document that evolves throughout the project lifecycle. Encourage regular updates to capture changes, additions, or modifications to the work breakdown structure.

4. **Training and Awareness:** Provide training and awareness sessions to project team members about the importance and usage of the WBS dictionary. Ensure that everyone understands how to access, interpret, and utilize the information effectively.

A well-implemented WBS dictionary is a powerful tool that enables project teams to effectively manage and communicate project scope, goals, and deliverables. By following the best practices outlined here, project teams can maximize the benefits of using a WBS dictionary, leading to improved project outcomes and stakeholder satisfaction.

CHAPTER 14

Case Studies: Real-world Examples

Construction Project: High-Rise Office Tower

As I have previously discussed, a WBS dictionary is an essential tool that can significantly contribute to the project's success. The construction of a high-rise office tower is an intricate and complex endeavor. The success of a project relies heavily on effective planning, organization, and execution. The idea is to give the project team insights into the importance of using a WBS dictionary for this sample project.

The WBSD elements selected are as follows:

- **WBS Code & Element:** The Work Breakdown Structure (WBS) is crucial for high-rise construction projects due to their complexity and scale. Each element of the construction, represented by a unique WBS code, allows for detailed planning and tracking. This granularity ensures that each phase, from foundation laying to interior finishing, is meticulously outlined and managed.

- **Facility & Equipment:** When constructing a high-rise office tower, managing the facility and equipment is a mammoth task. By breaking down these needs in the WBS, the project team can effectively schedule cranes, concrete mixers, and other heavy machinery, ensuring that these resources are available and optimally used when needed.

- **Base Cost, Contingency, Total Cost:** Financial control is paramount in large-scale construction. The WBS aids in estimating the base cost of each project element while also accounting for contingencies (see Note) - unforeseen expenses that are common in construction.

The WBS Approach

- **Detailed Breakdown and Assignment:** The project is divided into manageable sections, each with a specific WBS code and element description. This detailed breakdown allows for assigning specific teams and resources to each part of the project, enhancing accountability and efficiency. The sample is a high-level summary.

- **Resource Allocation and Scheduling:** By identifying facility and equipment needs in the WBS, the project manager can schedule their use effectively, avoiding bottlenecks with resources (people and equipment) and ensuring smooth progress during execution.

- **Budgeting and Financial Management:** The WBS provides a framework to estimate and track costs at each project stage. It helps create a comprehensive budget, including a contingency fund to manage risks and unexpected expenses.

Controlling the WBSD

- **Project Initiation and Planning:** A WBS framework is developed per the project's scope and structure. This early-stage implementation sets the foundation for all subsequent planning and execution activities.

- **Throughout Project Execution:** The WBS is a living document, continually referenced and updated throughout the project lifecycle. It guides daily operations, resource allocation, progress tracking & monitoring & control.

- **In Response to Changes and Challenges:** In construction, changes happen, and the WBS structure allows for flexible responses to these changes, providing a framework to assess impacts on time, cost, and resource requirements.

By implementing the WBS approach, the project can be managed with greater precision, efficiency, and control, significantly increasing its chances of successful completion within budget and on schedule.

WBS Code	Element	Facility	Equipment	Base Cost	Duration (weeks)	Contingency (%)	Total Cost
1	Site Preparation	Construction Site	Excavators, Bulldozers	$500,000	4	10	$550,000
2	Foundation Work	Building Base	Concrete Mixers, Cranes	$2,000,000	8	15	$2,300,000
3	Structural Framework	Building Frame	Cranes, Welding Machines	$5,000,000	20	15	$5,750,000
4	Exterior Construction	Building Exterior	Scaffolding, Lifts	$3,000,000	16	10	$3,300,000
5	Interior Construction	Building Interior	Paint Sprayers, Drills	$4,000,000	24	12	$4,480,000
6	Electrical Systems	Building Interior	Electrical Tools	$1,500,000	12	10	$1,650,000
7	Plumbing & HVAC	Building Interior	Pipes, HVAC Systems	$2,000,000	12	15	$2,300,000
8	Final Touches	Entire Building	Cleaning Supplies	$500,000	4	5	$525,000
9	Inspection & Handover	Entire Building	Inspection Tools	$250,000	2	5	$262,500

Table 11 – Sample Construction Project Basic WBSD

The table above shows a very simplified WBS structure. We establish a detailed Work Breakdown Structure Dictionary (WBSD) in Appendix 1, describing and explaining each element.

NOTE: *Keep the KISS principle in mind when putting a WBSD together!*

It is crucial to manage contingency separately and not form part of the control budget.

You do not need to control and monitor as many elements as shown in the appendix.

Software Development Project

This Software Development Project focuses on creating a new application tailored to meet specific business needs. The project encompasses various stages, from initial concept development to final deployment and maintenance. The emphasis is on clear and well-defined elements, comprehensive descriptions, tangible deliverables, and adherence to strict timelines and technical specifications.

Critical Aspects

- **Element and Description:** Each project stage, such as requirement gathering, design, coding, testing, and deployment, is clearly defined. Detailed descriptions clarify objectives, methodologies, and expected outcomes at each phase.

- **Deliverable:** The project comprises key deliverables, including design documents, code repositories, test reports, and the final software product. These deliverables are crucial milestones, marking the completion of various project phases.

- **Duration (Pessimistic, Most Likely, Optimistic):** Time management is critical in software development. Each phase has three estimated durations: Pessimistic (P), accounting for potential delays; Most Likely (ML), the expected realistic duration; and Optimistic (O), the best-case scenario. This approach helps in effective planning and risk mitigation.

- **Technical References:** The project adheres to high technical standards, utilizing best practices, coding standards, and industry guidelines. These references ensure the software is robust, scalable, and meets quality benchmarks.

- **Risk Code:** Potential risks, such as technological challenges, scope creep, or resource limitations, are identified and coded. This coding system helps monitor, manage, and mitigate risks during the stages.

WBS Code	Element	Description	Deliverable	Duration (P, ML, O)	Responsible
1	Project Management	Overall management of the project	Project plan, reports	2w, 1w, 1w	Project Manager
1.1	Requirement Gathering	Collecting user and system requirements	Requirements document	4w, 3w, 2w	Business Analyst
1.1.1	User Requirements	Gathering user-specific needs	User requirement specs	2w, 1.5w, 1w	Business Analyst
1.1.2	System Requirements	Defining system functionalities	System requirement specs	2w, 1.5w, 1w	Systems Analyst
1.2	Design	Designing the software architecture	Design documents	6w, 4w, 3w	Lead Developer
1.2.1	UI/UX Design	Designing user interface and experience	UI/UX prototypes	3w, 2w, 1.5w	UI/UX Designer
1.2.2	Database Design	Structuring the database components	Database design schema	3w, 2w, 1.5w	Database Architect
1.3	Development	Coding and building the software	Software build	12w, 8w, 6w	Development Team
1.3.1	Front-end Development	Developing the client-side interface	Front-end code	6w, 4w, 3w	Front-end Developers
1.3.2	Back-end Development	Developing server-side logic	Back-end code	6w, 4w, 3w	Back-end Developers
1.4	Testing	Verifying the software functionality	Test reports, bug reports	4w, 3w, 2w	QA Team
1.4.1	Unit Testing	Testing individual units of code	Unit test reports	2w, 1.5w, 1w	QA Engineers
1.4.2	Integration Testing	Testing combined components	Integration test reports	2w, 1.5w, 1w	QA Engineers
1.5	Deployment	Deploying the software in the production environment	Deployed application	2w, 1w, 0.5w	DevOps Team
1.6	Maintenance and Support	Post-deployment support and updates	Support logs, update patches	Ongoing	Support Team

Table 12 – Sample Software Project WBSD

The rationale for Emphasis

- Focusing on specific elements and their detailed descriptions in software development ensures that all team members and stakeholders understand the project's scope and objectives.

- Deliverables are the tangible outcomes that mark the progress and success of the project. They provide concrete goals for the team and measurable results for stakeholders.

- Estimating durations under different scenarios allows for flexible and realistic planning, catering to the dynamic nature of software projects.

- Adherence to technical references ensures the development a high-quality product that meets industry standards.

- Identifying and coding risks enables proactive risk management, a critical aspect of ensuring project success for a software project.

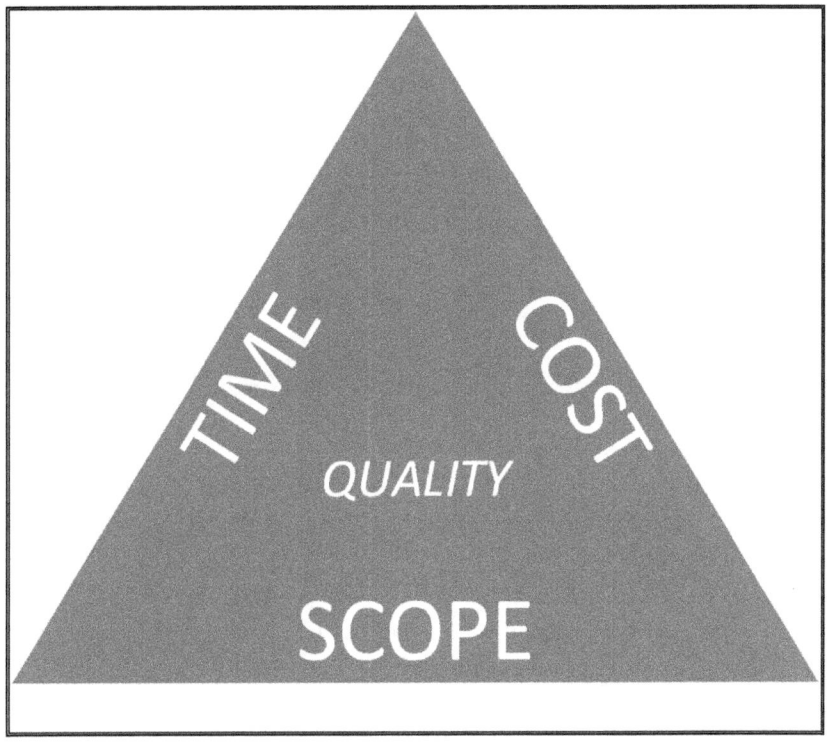

Figure 21 -Project Tri-Angle Scope, Cost, and Time (Quality)

This structured approach, focusing on these key aspects, is designed to deliver a high-quality software product within the planned timeline of the durations provided.

NOTE: *This project example has no resources and cost allocated, which every project must have to conform to the basic definition of a project.*

Marketing Campaign Project

A Marketing Campaign Project aims to create and execute a dynamic marketing strategy to enhance brand visibility and engagement. It involves coordinated marketing activities designed to target specific audiences and achieve predefined goals. The emphasis is on clearly defined project elements, detailed descriptions, measurable deliverables, tight schedule adherence, and clear responsibility assignments.

Key Aspects

- **Element and Description:** Each campaign component, such as market research, content creation, digital advertising, and public relations, is meticulously outlined. Descriptions provide detailed insights into each element's purpose, target audience, and intended message.

- **Deliverable Measurement:** Success metrics for each deliverable, such as engagement, lead generation, and conversion rates, are identified. This quantifiable approach ensures that the campaign's impact is measurable and aligns with the strategic objectives.

- **Duration:** The campaign is segmented into phases with specific timelines, from the initial planning stage to the execution and post-campaign analysis. Strict adherence to these timelines is crucial for the campaign's success, especially in fast-paced market environments.

- **Responsible:** Clear responsibility assignments for each element of the campaign are made. This includes assigning team members or departments to tasks such as content creation, social media management, and data analysis, ensuring accountability and efficient workflow.

The rationale for Emphasis

- Focusing on distinct elements with comprehensive descriptions ensures that the campaign addresses all necessary facets of marketing, from creative development to audience targeting.

- Measurable deliverables are critical in marketing, allowing the team to evaluate the effectiveness of each campaign element and make data-driven decisions.

- Given the dynamic nature of the marketing industry, managing tight schedules is essential to capitalize on market trends and consumer behavior.

- Assigning clear responsibilities ensures that every aspect of the campaign is managed by individuals or teams with the appropriate skills and expertise, enhancing the campaign's overall effectiveness.

By emphasizing these key aspects, the Marketing Campaign Project is designed to create a high-impact, results-driven marketing initiative that resonates with the target audience, achieves strategic goals, and enhances the brand's market presence within a well-structured and efficiently managed.

CHAPTER 15

WBSD - Conclusion and Next Steps

Recap of the Importance of Using a WBSD

As we discussed earlier, a WBS dictionary is an essential tool providing clarity, consistency, and efficiency to project teams.

First and foremost, a WBS dictionary acts as a common language for project teams. It ensures that everyone involved in the project understands the various components and tasks within the WBS. This common understanding eliminates confusion, misunderstandings, and potential errors that can arise when team members interpret the WBS differently. Using a standardized dictionary allows project teams to communicate effectively, streamline collaboration, and reduce the risk of miscommunication.

Furthermore, a WBS dictionary promotes consistency across projects. When multiple projects are executed within an organization, having a standardized dictionary ensures that similar tasks and deliverables are defined consistently across all projects. This consistency facilitates knowledge transfer, improves project coordination, and enhances the overall efficiency of the project management process. It also enables project managers to leverage best practices and lessons learned from previous projects, improving project outcomes.

Another key benefit of using a WBS dictionary is that it enhances project planning and estimation. By providing detailed descriptions, dependencies, and resource requirements for each WBS component, the dictionary enables project teams to develop accurate project schedules, allocate resources effectively, and estimate project costs more precisely. This level of granularity ensures that project plans are realistic and achievable, reducing the chances of project delays, scope creep, and cost overruns.

Finally, a WBS dictionary serves as a valuable reference tool throughout the project lifecycle. It provides a comprehensive overview of the project's scope, objectives, and deliverables, making it easier for project teams to monitor progress, track milestones, and evaluate project performance. Additionally, the dictionary helps identify potential risks, bottlenecks, and dependencies, allowing project managers to proactively address issues and make informed decisions.

In conclusion, the importance of using a WBS dictionary cannot be emphasized enough. It serves as a common language, promotes consistency, enhances planning and estimation, and provides a valuable reference tool for project teams. By implementing a WBS dictionary, project teams can improve communication, streamline collaboration, and ultimately deliver successful projects within budget and on time.

Key Takeaways for Project Teams

Here are some key takeaways for project teams regarding the importance of using a WBS dictionary in their day-to-day activities.

Firstly, a WBS dictionary serves as a comprehensive reference guide that defines and explains all the terms and elements used in a project's WBS. By having a common understanding of terminology, project teams can avoid confusion and ensure that everyone is on the same page. This is especially important when working with cross-functional teams or stakeholders from different departments or even external partners.

Secondly, the WBS dictionary helps to establish consistency and standardization across projects. It provides a framework that can be applied consistently throughout an organization, facilitating collaboration and knowledge sharing. With a common language and structure, project teams can easily communicate project goals, objectives, and deliverables, ensuring that everyone is aligned and working towards the same objectives.

Moreover, the WBS dictionary enhances project planning and estimation accuracy. By breaking down the project into smaller, manageable components, teams can better estimate resources, timeframes, and costs associated with each task. This results in more accurate project planning, reducing the risk of delays, budget overruns, and scope creep.

Furthermore, the WBS dictionary promotes effective risk management. By clearly defining the scope and boundaries of each deliverable, potential risks and dependencies can be identified early on. This allows project teams to proactively manage risks, mitigate potential issues, and develop contingency plans, ultimately increasing the chances of project success.

Lastly, the WBS dictionary serves as a valuable tool for project documentation and knowledge retention. It provides a centralized repository of project information, making it easier for team members to access and refer to past projects, lessons learned, and best practices. This promotes continuous improvement and enhances future project planning and execution.

In conclusion, a WBS dictionary is an essential resource for project teams. It promotes clarity, consistency, and collaboration, resulting in more effective project management. By using a WBS dictionary, project teams can streamline their work, enhance communication, and ultimately increase the likelihood of project success.

Next Steps for Implementing a WBSD

Once you understand the importance of using a Work Breakdown Structure (WBS) dictionary, it's time to take the next steps in implementing this

210

valuable tool within your project team. A WBS dictionary serves as a comprehensive reference guide that defines and describes each element of your project's WBS. It provides clarity, consistency, and a common language for all team members involved.

To successfully implement a WBS dictionary, it is crucial to follow a few key steps. First and foremost, you need to establish a clear and structured format for documenting the WBS dictionary. This format should include essential information such as the WBS element name, its identification number, a detailed description, associated deliverables, responsible party, and any dependencies or constraints.

Next, you should collaborate with your project team to gather the necessary information for the WBS dictionary. This step involves engaging subject matter experts and stakeholders to ensure that each element is accurately defined and described. Encourage open communication and invite feedback to ensure the completeness and accuracy of the information provided.

Once you have gathered all the relevant data, it's time to document and organize it within the WBS dictionary. You can use various tools and software to create a user-friendly and easily searchable format. Consider incorporating visualization techniques, such as flowcharts or diagrams, to enhance the understanding and accessibility of the information.

After completing the initial documentation, it's essential to continually update and maintain the WBS dictionary throughout the project lifecycle. As your project progresses and evolves, new elements may be added or existing ones may require modification. Regularly review and validate the dictionary to ensure it remains up-to-date and aligned with the project's objectives.

Furthermore, ensure that all project team members are aware of the WBS dictionary and its importance. Conduct training sessions to familiarize them with the tool and its benefits. Encourage consistent usage and emphasize its role in promoting effective communication, minimizing misunderstandings, and facilitating collaboration among team members.

In conclusion, implementing a WBS dictionary is critical to project success. By following these next steps, you can ensure that your project team has a comprehensive and reliable reference guide that enhances communication, reduces ambiguity, and improves overall project performance. Embrace the power of a WBS dictionary and reap the benefits of a well-structured project management approach.

Resources and References

A WBS dictionary serves as a comprehensive reference guide that provides detailed information about the components of a project's WBS. It acts as a repository of all the project's tasks, deliverables, milestones, and dependencies. By utilizing a WBS dictionary, project teams can streamline their communication, enhance collaboration, and ensure a common understanding of the project's scope and objectives.

One of the key benefits of using a WBS dictionary is that it promotes consistency and accuracy throughout the project. It serves as a single source of truth for the project team, enabling them to refer to standardized terminology and definitions. This consistency helps eliminate confusion, reduce rework, and ensure that everyone is on the same page.

Furthermore, a WBS dictionary provides a valuable resource for onboarding new team members. When new members join a project, they often face the challenge of understanding the project's structure, tasks, and deliverables. By providing them with access to the WBS dictionary, the learning curve can be significantly reduced. New team members can quickly familiarize themselves with the project's components, enabling them to contribute effectively from day one.

Another advantage of using a WBS dictionary is the ability to track and manage project changes. As projects evolve, it is common for the scope to expand, requirements to change, or new deliverables to emerge. The WBS dictionary acts as a reference point to capture these changes accurately. It allows project teams to update the dictionary with the latest information, ensuring that everyone is aware of the modifications and their impact on the project's overall structure.

Additionally, a WBS dictionary can serve as a valuable resource for future projects. By documenting the project's WBS components, lessons learned, and best practices, project teams can build a knowledge base that can be leveraged for future endeavours. This knowledge transfer ensures that the project team's experiences are not lost and can be used as a reference for similar projects in the future.

In conclusion, a WBS dictionary is an invaluable resource for project teams. It promotes consistency, aids in onboarding new team members, facilitates effective change management, and creates a knowledge base for future projects. By utilizing a WBS dictionary, project teams can enhance their communication, collaboration, and overall project success.

NOTE: And finally, let me say it again!

Keep it Simple, Simon (KISS) (or is it Stupid).

Always keep the KISS principle in mind.

PART III – REVIEW PRACTICES

CHAPTER 16

Develop a WBS

Follow these WBS Steps

Project Work Breakdown Structure (WBS) Sequence development. The objective is to produce a comprehensive, clearly defined, and thoroughly validated Work Breakdown Structure (WBS) that serves as a roadmap for the entire project.

The sequence of activities to achieve this is as follows:

1. **Initial Project Scoping**

   - Consult the project charter to understand and define the project's objectives, scope, and key stakeholders.

   - Document these elements for future reference.

2. **Gather Requirements**

   - Conduct stakeholder interviews, surveys, or workshops.

   - Document the findings, focusing on the project's technical, functional, and business requirements.

3. **Conduct Preliminary Research**

   - Research industry best practices for similar projects.

   - Note any assumptions or constraints that will impact the project.

4. **Identify Major Deliverables**

   - Generate a list of significant outcomes or deliverables.

   - Validate this list with stakeholders and project sponsors.

5. **Initial WBS Drafting**

   - Decompose major deliverables into smaller, manageable tasks and sub-tasks.

   - Utilize a WBS software or a hierarchical chart to visualize this breakdown.

6. **Assign Responsibility**

   - Designate task ownership to specific team members or roles.

   - Ensure each task and sub-task is assigned to someone capable.

7. **Estimate Time and Cost**

   - Estimate the time and resources needed for each task.

   - Record these estimates for integration into the project plan.

8. **WBS Validation**

   - Conduct a review session involving stakeholders to validate the drafted WBS.

   - Implement feedback and revise as necessary.

9. **Quality Assurance**

   - Use an internal checklist to review the WBS for adherence to quality standards.

   - Make adjustments as required.

10. **Finalize and Distribute WBS**

   - Get final approvals on the WBS.

   - Distribute to all stakeholders, including team members and sponsors.

Activity Instructions

Objective

To create a detailed, validated, and high-quality Work Breakdown Structure (WBS) that effectively guides the project to successful completion.

Instructions

1. Follow the sequence of activities in the WBS Sequence scrupulously.
2. Keep documentation for each step for traceability and auditing.
3. Utilize the Self-Check Mechanism for quality assurance at each step.

Deliverables

- Project Scope Document
- Requirements Documentation
- Initial and Final WBS
- Time and Cost Estimates
- Validation and Quality Assurance (QA) Reports

Grading Criteria for a Review

- Completeness: 40%
- Stakeholder Engagement: 20%
- Technical Accuracy: 20%
- Timeliness: 10%
- Quality Assurance: 10%

Review Submission Guidelines

Teams must submit all required deliverables as a single, organised package through the designated document control project management software (if one is used by the project team).

Potential Review Outcomes

- Full Success: Fully aligned WBS.

- Partial Success: Minor adjustments required.

- Failure: Major realignment or potential project failure due to incorrect WBS setup.

Self-Check Mechanism Table

Activity Steps	Check Criteria	Verification Mechanism	Completed (Yes/No)
Initial Project Scoping	Objectives align with charter	Review against project charter	
Gather Requirements	All requirements are documented	Stakeholder interviews	
Preliminary Research	Assumptions and constraints noted	Documented review	
Identify Deliverables	All major deliverables listed	Stakeholder validation	
Initial WBS Draft	Includes all tasks and sub-tasks	Internal review	
Assign Responsibility	Each task has an owner	Task ownership list	
Estimate Time and Cost	Estimates are realistic	Cross-reference with past projects	
WBS Validation	Stakeholder approval obtained	Stakeholder sign-off	
Quality Assurance	Adherence to quality standards	QA checklist	
Finalize and Distribute	WBS distributed to all	Email confirmation or documentation	

Table 13 – Self-check for Developing a WBS

Using the Self-Check Mechanism Table, each team member should know what they are responsible for and what criteria must be met to complete each step. This empowers the team to independently validate the quality and completeness of their work at each stage, thereby maintaining the project's integrity.

CHAPTER 17

Develop a Bottom-up Estimate

Follow this sequence of activities to deliver a proper bottom-up estimate. This is the best way to provide a detailed estimate for your project. You could also use other methods, but I recommend a bottom-up estimate.

To deliver a proper bottom-up estimate, follow this sequence of activities:

1. **Define the Scope:** Clearly define the scope of the project or task to be estimated. This includes understanding the objectives, deliverables, and any constraints or assumptions that may affect the estimate.

2. **Breakdown Work**: Break down the project or task into smaller work packages or activities. This involves identifying all the individual tasks or components required to complete the project.

3. **Identify Resources:** Determine the resources needed for each work package or activity. This includes identifying the specific skills, materials, equipment, and any external resources required.

4. **Estimate Effort:** Estimate the effort required for each work package or activity. This involves determining the amount of time or effort needed to complete each task, considering factors such as complexity, dependencies, and any historical data or benchmarks.

5. **Estimate Duration:** Calculate the duration or timeline for each work package or activity. This involves considering the availability and allocation of resources, dependencies between tasks, and any potential delays or constraints.

6. **Estimate Costs:** Estimate the costs associated with each work package or activity. This includes considering the labor costs, material costs, equipment costs, and any other relevant expenses.

7. **Validate Estimates:** Review and validate the estimates with subject matter experts, stakeholders, or team members who have experience or expertise in the specific tasks or activities. This helps to ensure the estimates are accurate and realistic.

8. **Summarize Estimates:** Summarize all the estimates for each work package or activity, including effort, duration, and costs. This provides an overall view of the project or task's estimated requirements.

9. **Document Assumptions:** Document any assumptions made during the estimation process. This includes assumptions about resource availability, task dependencies, or any other factors that may impact the estimates.

10. **Review and Refine:** Review the estimates and refine them if necessary. This may involve making adjustments based on feedback, new information, or changes in project requirements.

11. **Present Estimate:** Present the bottom-up estimate to the relevant stakeholders or decision-makers. This includes explaining the methodology used, highlighting any key assumptions or risks, and providing a clear breakdown of the estimated costs, effort, and timeline.

12. **Update and Track:** Continuously update and track the estimate throughout the project lifecycle. This involves monitoring progress, comparing actuals against estimates, and making adjustments as needed.

By following this basic list, you can deliver a bottom-up estimate that is detailed, accurate, and aligned with the specific requirements of the project.

Activity Instructions

Objective:

The primary objective of this activity is to deliver an accurate and comprehensive bottom-up estimate for a given project or task. This estimate will provide detailed insights on the expected effort, duration, costs, and resources needed to complete the project or task successfully. It will serve as a baseline for planning, scheduling, budgeting, and tracking the project's progress.

Instructions:

Follow the sequence of activities as outlined. Start by defining the project scope and breaking down the work into manageable parts. Determine the resources and effort needed for each task, then calculate the estimated duration and costs. Validate your estimates with experts, summarize them, and document any assumptions made in the process. Review, refine, and present your estimate to stakeholders, then ensure it's updated and tracked throughout the project lifecycle.

Deliverables:

1. A detailed work breakdown structure (WBS) which includes all the individual tasks or components of the project.

2. A list of necessary resources for each task.

3. Detailed effort, duration, and cost estimates for each task.

4. A validation report containing feedback from experts or experienced team members.

5. A document outlining all assumptions made during the estimation process.

6. Finalized and summarized project estimate.

7. A plan for how the estimate will be reviewed, updated, and tracked throughout the project lifecycle.

Grading Criteria:

The quality of the bottom-up estimate will be evaluated on the following criteria:

1. Detail and accuracy of the work breakdown structure.

2. Thoroughness and accuracy of resource identification.

3. Reasonableness and accuracy of effort, duration, and cost estimates.

4. Quality of validation and the incorporation of expert feedback.

5. Clarity and relevance of documented assumptions.

6. Overall comprehensiveness and coherence of the final estimate.

7. Quality of the review and update plan for the estimate.

Submission Guidelines:

All deliverables should be compiled into one comprehensive report with clear headings and sections for each part. Any supporting documents, such as the WBS or resource lists, should be included as appendices.

Potential Outcomes:

By delivering a thorough and accurate bottom-up estimate, the project or task will have a clear roadmap for execution. This can lead to improved project planning, scheduling, and resource allocation. It can also aid in effective budgeting and cost control. On the other hand, if the estimate is inaccurate or lacks detail, it could lead to project delays, cost overruns, or resource shortages. Regular updates and tracking of the estimate will help mitigate these risks.

Self-Check Mechanism

Criteria	How to Check	Completed (Yes/No)
Completeness of Estimates	Ensure all work packages are included in the estimate.	
Accuracy	Validate calculations and cross-verify unit costs.	
Stakeholder Validation	Secure written or verbal approval from stakeholders.	
Documentation Quality	Check for clarity, organization, and completeness.	
Initial WBS Draft	Includes all tasks and sub-tasks	
Assign Responsibility	Each task has an owner	
Estimate Time and Cost	Estimates are realistic	
WBS Validation	Stakeholder approval obtained	
Quality Assurance	Adherence to quality standards	
Finalize and Distribute	WBS distributed to all	

Table 14 – Self-check for Bottom-up Estimate WBS

By following these guidelines and mechanisms, the team should be able to not only develop a detailed and robust Bottom-up Estimate but also ensure that it meets all quality and stakeholder expectations.

CHAPTER 18

Develop a WBS Dictionary

Follow these WBS Steps

Project Work Breakdown Structure (WBS) Sequence development. The objective is to produce a comprehensive, clearly defined, and thoroughly validated Work Breakdown Structure (WBS) that serves as a roadmap for the entire project. The sequence of activities to achieve this is as follows:

1. **Work Package Identification**

   - Start by identifying the work packages from the WBS that require detailed descriptions in the dictionary.

2. **Scope Clarification**

   - Review the scope of each work package and clarify any ambiguities or uncertainties with project stakeholders.

3. **Detailed Descriptions**

   - Create detailed descriptions for each work package, including its scope, objectives, deliverables, and dependencies.

4. **Resource Requirements**

   - Specify the resources needed for each work package, including labor, materials, equipment, and subcontractors.

5. **Cost Estimates**

   - Provide cost estimates for each work package, detailing the budget allocated to complete the work.

6. Schedule Information

- Include scheduling information, such as start and end dates, milestones, and critical path details for each work package.

7. Risk Assessment

- Identify potential risks associated with each work package and describe mitigation strategies.

8. Quality Standards

- Define the quality standards and criteria that must be met to complete each work package successfully.

Activity Instructions

Objective:

This activity aims to create a comprehensive WBS Dictionary that provides detailed descriptions of each work package in the project.

Instructions:

Work Package Identification: Identify which work packages from the WBS require detailed descriptions in the dictionary.

Scope Clarification: Review the scope of each work package and seek clarification from stakeholders if needed.

Detailed Descriptions: Create comprehensive descriptions for each work package.

Resource Requirements: Specify the necessary resources for each work package.

Cost Estimates: Provide cost estimates for each work package.

Schedule Information: Include scheduling details for each work package.

Risk Assessment: Identify and document potential risks for each work package.

Quality Standards: Define quality standards for each work package.

Deliverables:

A completed WBS Dictionary with detailed descriptions for each work package.

Review Grading Criteria:

Work Package Identification: Correct identification of work packages (10 points).

Scope Clarification: Effective clarification of scope with stakeholders (10 points).

Detailed Descriptions: Comprehensive and clear descriptions (15 points).

Resource Requirements: Accurate resource specifications (10 points).

Cost Estimates: Precise cost estimates (10 points).

Schedule Information: Inclusion of relevant scheduling details (10 points).

Risk Assessment: Thorough identification of risks and mitigation strategies (15 points).

Quality Standards: Well-defined quality standards (10 points).

Submission Guidelines:

Submit the completed WBS Dictionary in electronic format and all deliverables should be compiled into one comprehensive report with clear headings and sections for each part. Any supporting documents, such as the WBS dictionary or glossary (if required).

Potential Outcomes:

Completing this activity will result in a detailed WBS Dictionary that comprehensively understands each work package within the project.

Self-Check Mechanism

Criteria	How to Check	Completed (Yes/No)
Work Package Identification	Verify that all relevant work packages are included.	
Scope Clarification	Ensure that any unclear scope aspects have been clarified.	
Detailed Descriptions	Check if descriptions are comprehensive and easy to understand.	
Resource Requirements	Confirm that all necessary resources are specified.	
Cost Estimates	Validate the accuracy of cost estimates.	
Schedule Information	Check if scheduling details are provided for each work package.	
Risk Assessment	Verify the presence of identified risks and mitigation plans.	
Quality Standards	Ensure that quality standards are well-defined.	

Table 15 – Self-check for Basic WBS Dictionary

Following the self-checklist will enable users or team members to develop a WBS dictionary that covers all requirements of your project. See a full example of dictionary elements that you could add to your project in the Real-world section and project examples.

CHAPTER 19

Project Office WBS Guideline (Example)

Purpose and Scope of the Guideline

The primary aim of a Work Breakdown Structure (WBS) is to break down the comprehensive scope of a project into distinct, measurable components that can be: Allocated resources, priced, scheduled, and managed effectively. The WBS serves as a unified communication platform throughout the product / project lifecycle.

Objectives

A. Best Practice Guideline for WBS:

Establish a systematic approach that serves as a universal best practice for constructing and applying a thorough WBS. This will aid in the evaluation and progression of projects through your organization's stage gate approval process (if you have any), particularly for major projects.

B. Benchmarking:

To compare major projects within various industries by applying a standardized WBS.

C. Supply Chain / Procurement Management:

A WBS must support your organization's Supply Chain / Procurement initiatives and should be aligned in delivering a successful project.

D. Estimate vs. Control Budget:

A WBS always starts with a project estimate that is aligned to the deliverables. The transition from a project estimate to a control budget must be aligned with the original scope and stakeholders. When stakeholders change, transitioning from an estimate to a control budget and keeping scope creep at bay is very difficult and requires additional communication with the stakeholders.

E. WBS Dictionary:

Project management can enhance efficiency and clarity by establishing a systematic approach as a universal best practice for creating and utilizing a comprehensive Work Breakdown Structure (WBS) Dictionary. This practice is particularly valuable for navigating your organization's stage gate approval process, especially in the context of major projects.

A well-structured WBS Dictionary is the key reference document, offering detailed descriptions, resource requirements, cost estimates, scheduling information, risk assessments, and quality standards for each work package. This guideline ensures stakeholders have a clear and unified understanding of project components, fostering effective decision-making and project success.

Summary of Contents

Project management is a holistic field where various elements are interconnected. In light of this, the document aims to provide a comprehensive guide to:

- Elaborate on the project scope in detail;
- Construct a Work Breakdown Structure (WBS) that integrates seamlessly with contemporary database systems for project estimation;
- Clarify the Scope of Work through the lens of the WBS;
- Identify the essential attributes of a WBS that facilitate effective project control estimates, aligned with Work Packages (WPs) that are actionable;

- Transition WPs into System Packages (SPs) to facilitate the commissioning, handover, and ramp-up phases.
- Have a comprehensive WBS Dictionary that is sufficient detail to control the project.

WBS with a Systems alignment:

Project Managers usually do not implement a SP approach to their WBS, which catches up with them later in the project. Have you ever heard of projects where the progress hovers around 95-98% for months on end? *This is the reason why!* No provision has been made for a systems approach in your WBS. I get into much more detail around this topic in my project scheduling & planning books.

Detailed Project Scope

Inputs

- Project Charter
- Initial Scope Statement
- Needs Assessment
- Historical Data from Similar Projects
- Process and Utility Flow Diagrams
- Equipment Inventories
- Site Layouts

Outputs

- Comprehensive Scope Statement
- Project Goals and Priorities
- Project Boundaries
- Deliverables
- Scope Management Plan
- Assumptions and Constraints
- Exclusions
- External Factors

Methodology

- Market/Product Analysis
- Selection of Optimal Alternatives
- Value-Improving Practices
- Expert Consultation
- Stakeholder Evaluation

Work Breakdown Structure (WBS)

Objectives

- To create a WBS that aligns the project scope with accurate estimates and schedules.
- To define client deliverables clearly.
- To establish WPs that translate the project estimate into a controllable budget.
- To allow for the reorganization of WPs into SPs for efficient commissioning and ramp-up.
- To facilitate the identification of scope changes.

Inputs

- **Project Estimation Requirements**
  - WBS Guidelines
  - Project Charter
  - Scope Statement
  - Equipment Inventories
  - Site Layouts
  - Process Flow Diagrams
- **Cost Control and Scheduling Requirements**
  - Scope Management Plan
  - Project Execution Plan

Outputs

- Hierarchical WBS
- Database fields for WBS decomposition

- Unique and separable WPs
- WBS Dictionary
- Sort codes for database manipulation
- Reconciliation of total estimate with budget

Methodology for WBS Preparation

- Selection of compatible database systems
- Determination of decomposition criteria
- Consistent detailing across categories
- Elimination of ambiguous naming
- Agreement on WBS elements
- Definition of entry-level elements

Work Package (WP) Development

- Assign WPs to the lowest WBS level.
- Ensure WP uniqueness.
- Align elements within each WP.
- Exclude incompatible elements.
- Assign WPs to specific responsibilities.
- Ensure compliance with the project strategy.
- Define WP start and finish dates, budget, and duration.

Conclusion

While this guide focuses on Studies (Feasibility), the principles are universally applicable across different phases of project development. Employing a consistent system and WBS from the Conceptual to the Pre-feasibility stages will streamline stage gate reviews and ensure project success. Ensure that your project delivers a WBS for execution that is as detailed as possible. Use the methods in this book to guide you!

Appendix: WBS Dictionary Templates

A Work Breakdown Structure (WBS) is a powerful tool that helps project teams plan and execute their projects effectively. It provides a hierarchical framework that breaks down the project scope into manageable and organized components. However, to fully utilize the benefits of a WBS, it is crucial to incorporate a WBS dictionary into your project management practices.

The importance of using a Work Breakdown Structure dictionary cannot be overstated. It serves as a comprehensive reference guide that defines and describes each component of the WBS. The WBS dictionary promotes effective communication and collaboration among project team members by providing a common language and understanding.

The basic WBS dictionary format is a simple yet effective way to document the components of your WBS. It consists of essential information such as the WBS component name, unique identifier, description, and responsible party. Let's take a closer look at each element:

1. **WBS Component Name:** This is the name or title given to each component of the WBS. It should be concise, descriptive, and reflect the work or deliverable it represents.

2. **Unique Identifier:** Assigning a unique identifier to each component helps in tracking and referencing them throughout the project lifecycle. It can be a combination of letters, numbers, or a coding system specific to your organization.

3. **Description:** The description provides a detailed explanation of what the component entails. It should be clear and specific, defining the scope and boundaries of the work to be completed.

4. **Responsible Party:** Identifying the person or team responsible for

executing each WBS component is essential. This ensures accountability and facilitates effective project coordination.

Heading	Description	Explanation
Level	Hierarchical stage in the WBS	Represents the stage or tier in the project structure.
WBS Code	Unique identifier for each element	Used for tracking and organizing project elements.
Element	Specific component of the project	Refers to individual tasks or deliverables.
Description	Detailed explanation of the element	Provides clarity on what each element entails.
Deliverable	Tangible or intangible output	The end product or service resulting from the element.
Acceptance Criteria	Standards for deliverable acceptance	Defines how a deliverable is deemed complete or acceptable.
Area	Geographic or logical segment	Specifies the location or segment of the project.
Sub Area	Subdivision of the area	Further breakdown of the project area.
Facility	Specific facility involved	Identifies the facility where work is conducted.
Equipment	Tools or machinery used	Lists the equipment required for the project.
Capital Code	Financial identifier	Used for budgeting and financial tracking.
Contract Package No.	Identifier for contractual agreements	Links elements to specific contracts.
Code of Accounts	Financial management tool	Organizes expenses and revenues.
Responsible	Party accountable for the element	Identifies who is in charge of each element.
Base Cost	Estimated cost without contingencies	The initial cost estimate of the element.
Duration (P, ML, O)	Estimated time frame	Duration in Pessimistic, Most Likely, and Optimistic scenarios.
Contingency	Extra resources for unexpected events	Allocated for unforeseen changes or challenges.
Escalation	Adjustment for inflation or cost increases	Accounts for rising costs over time.
FOREX	Foreign exchange considerations	Relevant for projects with international components.
Total Cost	All-inclusive cost estimate	Sum of all expenses associated with the element.
Technical References	Relevant standards or guidelines	Technical documents or standards guiding the work.
Deliverable Measurement	Metrics for deliverable assessment	How the success of a deliverable is measured.
Risk Code	Identifier for potential risks	Classifies and tracks different risks.
Discipline	Specialized area of expertise	Identifies the professional field relevant to the element.

Table 16 - WBS Dictionary Description and Explanation

By using the basic WBS dictionary format, project teams can easily access and understand the details of each component within the WBS. This clarity helps in accurate planning, estimation, and resource allocation. It also aids in identifying dependencies and interrelationships between different components, enabling efficient project scheduling.

Moreover, a WBS dictionary serves as a valuable reference tool for future projects. As organizations undertake similar projects or engage in continuous improvement initiatives, the WBS dictionary can be used as a knowledge base to leverage past experiences, lessons learned, and best practices.

By incorporating a WBS dictionary into your project management practices is crucial for project success. The basic WBS dictionary format provides a structured and standardized approach to documenting and communicating each component's details within the WBS. A WBS dictionary helps project teams plan, execute, and deliver their projects more efficiently and accurately by promoting clarity, understanding, and effective communication.

References:

Burke, R 1990, *Project management: planning and control*, Unique Press (Pty) Ltd, Montague Gardens, Cape.

Burke, R 2007, *Project management: techniques*, college edn, Burke Publishing, Everbest, HK, China.

Creese, RC & Fang, Y 2010, 'Time Based Schedule Performance Index', *Cost Engineering*, 3 March 2010, vol. 52, no. 3, pp. 18-20.

Dingle, J & Jashapara, A 1995, *The commercial project manager: managing owners, sponsors, partners, supporters, stakeholders, contractors and consultants*, McGraw Hill, London, pp. 76–99.

Gardiner, P 2005, *Project management: a strategic planning approach*, Palgrave Macmillan, Basingstoke, UK.

Globerson, S 1994, 'Impact of various work-breakdown structures on project conceptualization', *International Journal of Project Management : the Journal of the International Project Management Association*, vol. 12, no. 3, pp. 165-171.

Griffith, AF 2005, Scheduling Practices and Project Success, PS.05, American Association of Cost Engineers (AACE), Morgantown WV, USA.

Hassen, N 1990, *Developing a creative and innovative project plan*

: an output oriented approach, Just Planning Pty Limited, Canberra, ACT, pp. 7.1-7.

Independent Project Analysis (IPA) 2006b, *Scheduling Practices Course: A Course for Project Managers*, The IPA Institute, Independent Project Analysis Inc, New York.

International Project Management Association (IPMA) 2006, *ICB- IPMA Competence Baseline*, 3rd edn, International Project Management Association, Nijkerk, The Netherlands.

Kerzner, H 1984, Project management: a systems approach to planning, scheduling and controlling, 2nd edn, Van Nostrand Reinhold Company Inc, New York.

Kiess, TE & Morgan, S 2010, 'Six States Defined by Earned Value Variance and Its Use to Form New Project Performance Indicators', *Cost Engineering*, 3 March 2010, vol. 52, no. 3, pp. 10-17.

Lock, D 1988, Project management, 4th edn, Gower Publishing Company, Brookefield, Vermont.

Mansuy, J 1991, 'Work breakdown structure: a simple tool for complex jobs', Cost Engineering, vol. 33, no. 12, Dec., pp. 15–18.

PMI 2021, *A Guide to the Project Management Body of Knowledge (PMBOK® Guide) and Standard for Project Management*, 7th edn, Project Management Institute, Newtown Square, Pennsylvania.

PMI 2017, *A Guide to the Project Management Body of Knowledge (PMBOK ® Guide)*, 6th edn, Project Management Institute, Newtown Square, Pennsylvania.

PMI 2019, *Practice Standard for WORK BREAKDOWN STRUCTURES*, 3rd edn, Project Management Institute, Newtown Square, Pennsylvania.

Raz, T & Globerson, S 1998, 'Effective Sizing and Content Definition of Work Packages', *Project Management Journal : Project Management Institute*, vol. 29, no. 4, pp. 17-23, Project Management Institute, Inc, Newtown Square, Pennsylvania.

Roberts, J 1999, 'Project sponsors reduce IT risks', *The Australian*, 24 08, p. 2.

About the Author

Ernest Lehman is a well respected expert in Project Management (PM) and Project Services (PS) with 43 years of experience that used the PMBOK® Guide, IPMA®, IPA®, APM®, and AIPM® guides, processes and methodologies. He has worked with leading companies in engineering, construction, and mining, successfully managing and revitalizing major projects while mentoring teams to success.

Holding dual master's degrees with two "Dean's Awards for Outstanding Academic Achievement," Ernest is known for his analytical skills and process-driven strategies that optimize costs, mitigate risks, and ensure quality outcomes.

In his tenure with Fortune 500 companies, Ernest has a proven track record of overseeing a portfolio of 800 projects (220 active per year), maintaining exceptional CPI and SPI metrics way above 95% (year-on year), achieved through lessons learned and hard-earned experience.

Ernest is committed to shaping the industry's future leaders through his insightful books, which serve as a gateway to professional growth and cover a unique blend of ethics, governance compliance, and innovative thinking.

He founded Ernest Lehman University, a leading online educational platform that embodies his principles of insightful leadership, competency development, active learning, mentoring, and coaching.

Join Ernest to enrich your professional skills, find inspiration for your career, and embark on a path toward excellence.

Other Books by Ernest D. Lehman

Estimating for Construction Projects: A Project Manager's Roadmap to Success

Confidence is Key: How to Nail Your Job Interview and Land Your Dream Job.

Strategic Cost Management: Navigating Project Budgets with Confidence.

Foundation for Success: The Power of a Strong Project Baseline.

The Power of Control: The Ultimate Guide to Project Management Estimation

Earned Value Management: A Practical Approach to Project Performance Tracking.

Navigating EPC FIDIC Contracts: Strategies for Project Services Success.

Coaching the Leaders of Tomorrow: A Comprehensive Guide for Project Professionals.

The Power of Integration: A Project Team Member's Handbook.

Contact us

Welcome to ELU!

Visit our website for more information on our programs, courses, or coaching and mentoring sessions that will help you accelerate your career.

What You Will Learn:

This book demystifies the Work Breakdown Structure (WBS) and introduces you to other project structures, such as: Organizational Breakdown Structure (OBS), Product Breakdown Structure (PBS), and Work Package Breakdown Structure (WPBS). The practical activity sheets will arm you with actionable strategies and insights.

- **Master the Basics:** Understand the fundamentals of WBS, its purpose, and benefits.
- **Diverse Structures:** Learn about other critical project structures like OBS, PBS, and WPBS and how they interlink with WBS.
- **Real-world Application:** Dive into practical examples and case studies that demonstrate the theories in action.
- **Step-by-Step Guidance:** Follow a detailed roadmap for creating a highly effective WBS, from defining scope to breaking down deliverables.
- **Best Practices:** Gain insights into the industry's best practices for constructing and utilizing work breakdown structures.
- **Reliable Activity Worksheets:** Leverage actionable worksheets designed to guide you through real-world applications of WBS and other structures.
- **Self-Assessment:** Take advantage of self-check mechanisms to evaluate your understanding and application of the material.

Whether you are a novice project manager or an experienced professional, this book offers a comprehensive, one-stop resource for work breakdown structures.

The key takeaway? You'll not just learn to organize projects; you'll master the art of breaking down projects into manageable pieces that set the stage for success.

Get ready to turn the page with lots of real samples and transform the way you approach project management.

www.ingramcontent.com/pod-product-compliance
Lightning Source LLC
Chambersburg PA
CBHW072141290526
45794CB00004B/1388